Interaction of Particles and Radiation with Matter

.

Springer
Berlin
Heidelberg
New York
Barcelona
Budapest
Hong Kong
London
Milan
Paris
Santa Clara
Singapore
Tokyo

Vsevolod V. Balashov

Interaction of Particles and Radiation with Matter

Translated by Gil Pontecorvo

With 119 Figures

 Springer

Professor Dr. Vsevolod V. Balashov

Moscow State University, Institute of Nuclear Physics
119899 Moscow, Russia

Translator

Dr. Gil Pontecorvo

Joint Institute for Nuclear Research
14198 Dubna, Moscow Region, Russia

Title of the original Russian edition: *Strojenie veshchestva*
© Moscow University Press 1993 · All rights reserved

ISBN-13: 978-3-642-64383-5   e-ISBN-13:978-3-642-60386-0

DOI: 10.1007/978-3-642-60386-0

Springer-Verlag Berlin Heidelberg NewYork
Library of Congress Cataloging-in-Publication Data.
Balashov, V. V. (Vsevolod Viacheslavovich) [Stroenie veshchestva. English] Interaction of particles and radiation with matter / Vsevolod V. Balashov; translated by Gil Pontecorvo. p. cm. Includes bibliographical references and index. ISBN 978-3-642-64383-5 (Berlin: acid-free paper)1. Matter–Effect of radiation on. 2. Particles (Nuclear physics) I. Title. QC173.39.B3513 1996 530.4'16–dc20 96-8637

© Springer-Verlag Berlin Heidelberg 1997
Softcover reprint of the hardcover 1st edition 1997

The use of general descriptive names, registered names, trademarks, etc. in this publication does not imply, even in the absence of a specific statement, that such names are exempt from the relevant protective laws and regulations and therefore free for general use.

Typesetting: Data conversion by Adam Leinz, Karlsruhe
Cover design: *design & production* GmbH, Heidelberg

SPIN 10478336   56/3144 – 5 4 3 2 1 0 – Printed on acid-free paper

Preface

I am very glad to encounter the new readers of my books. It is published without any changes or supplements to the Russian text, even though many new results relevant to the issues raised have been obtained in the laboratories of various countries. I do hope this edition, which retains all the aspects of the original, will turn out to be useful not only for students just entering the physics of the microworld, but also for their professors as an example of how specialists in physics are educated in the universities of my country.

I am grateful to Dr. G. Pontecorvo for undertaking the work of preparing the English translation of the book.

Moscow, November 1996   *V. V. Balashov*

Preface to the Russian Edition

This book is presented to the reader as the first part of a series of textbooks conceived by the author under a common title, "The structure of matter". The second part is to deal with applications of the methods of collision theory in atomic and nuclear physics, and the third will be devoted to the structure of nuclear matter. The text is based on a year-long course of lectures delivered for several years at the nuclear physics department of the Physical Faculty of the M. V. Lomonosov Moscow State University. The main purpose of this textbook is to serve as an "introduction to the profession" for future experimentalists and theorists who intend to work in the physics of the microworld. The text is intended for students in their sixth or seventh semester who are already familiar with the fundamentals of atomic physics. In parallel with this course, the students will presumably study quantum mechanics in accordance with the standard university program.

Issues of particle and radiation interactions with matter, dealt with in this book, are more often than not included in university programs as complementary, required only for further studies of experimental methods in atomic and nuclear physics and for work in laboratories. During recent years, however, we have witnessed a rise in interest in independent studies of physical processes accompanying the interaction of charged particles, electromagnetic radiation, and neutrons with matter. This fact has led us to diverge somewhat from traditional exposition of the subject, to weaken the accent on the presentation of the final results of studies, computations, tables, etc. and, on the contrary, to make the reader take as much part as possible in analyzing the physical essence of the phenomena dealt with. Experience reveals that such an analysis of the issues of particle and radiation interaction with matter turns out to be an excellent school for the physicist just starting in his or her career. The extreme clarity of the phenomena studied herein, the close mutual penetration of concepts relevant to the physics of the micro- and macro-worlds, the natural entanglement of the methods of classical and quantum theory applied, all serve to develop in the students physical intuition and broad views, so necessary for their future independent work. The reader will find in the book interesting examples of concrete results, taken from most recent experimental and theoretical publications. It must only be noted that the choice of illustrative material is based only on pedagogical arguments of clarity, and references to publications given in figure captions are not intended as a demonstration of their role in the establishment or of the development of the relevant fields of research.

To conclude: a few words about the title of the whole course. It reflects not just the straightforward meaning. "The structure of matter" was precisely the title of the whole set of lectures, seminars, and laboratory exercises in the physics of the micro-world, taught in the 1950s to us, Moscow State University students in nuclear physics, by V. I. Veksler, L. V. Groshev, I. M. Frank, D. I. Blokhintsev, Yu. M. Shirokov, and V. A. Petukhov. I dedicate this book to my teachers.

I am grateful to my colleagues N. G. Goncharova, F. A. Zhivopistsev, N. M. Kabachnik, and A. S. Yarov, who directed practical studies in this course, for numerous discussions.

Moscow, January 1993   *V. V. Balashov*

Contents

Part I. Interaction of Charged Particles with Matter

Part II. Interaction of Electromagnetic Radiation with Matter

Part III. Interaction of Neutrons with Matter

Part IV. Mesoatomic and Mesomolecular Processes

Introduction:
Fundamentals of the Structure of Matter

A. Elementary Particles

In the past decades, the most significant achievement of physics was the establishment of a small number of fundamental principles, on the basis of which an extremely rigorous and elegant classification of elementary particles was realized. New data on elementary particles, including those relevant both to newly revealed particles and to corrections of quantitative characteristics of already known particles, are regularly published in scientific journals. We shall present here a part of such data from tables published in 1990 [A.1].

The first group is composed of the gauge bosons (Table A.1). These include the photon, γ, and the intermediate bosons: the charged $W^{\pm}$ and neutral Z^0 bosons. The photon serves as the carrier of electromagnetic interactions and the W and Z bosons carry weak interactions. They are all vector particles, since their intrinsic angular momentum is unity, and their intrinsic parity is negative ($J^P = 1^-$).

Table A.1. Properties of gauge bosons ($J = 1$)

Particle	Mass m /MeV	Decay width Γ /GeV	Decay channels
γ	$< 3 \times 10^{-33}$	stable	
$W^{\pm}$	$80\,600 \pm 400$	2.25 ± 0.14	$e\nu$ ($\sim 10\,\%$) $\mu\nu$ ($\sim 10\,\%$) $\tau\nu$ ($\sim 10\,\%$)
Z	$91\,161 \pm 31$	2.534 ± 0.02	hadrons ($\sim 70\,\%$) e^+e^- ($\sim 3\,\%$) $\mu^+\mu^-$ ($\sim 3\,\%$) $\tau^+\tau^-$ ($\sim 3\,\%$) $\nu\bar{\nu}$ ($\sim 19\,\%$)

The theory unifying electromagnetic and weak interactions combines the photon and the intermediate bosons in a sole group. There also exists an essential difference between them: the photon is a massless particle, whereas the W and Z bosons are the heaviest known elementary particles. There is here a straightforward relation to the electromagnetic interaction being long ranged, while the weak interaction is the shortest ranged. Particle masses together with their mean lifetimes are presented in Table A.1. According to the "uncertainty principle" relating lifetime and energy, the width of the spread in energy (mass) of a compound may serve as an equivalent of its mean lifetime:

$$\Gamma = \frac{\hbar}{\tau} \, . \tag{A.1}$$

In Tables A.1–A.4, the values of Γ, instead of τ, are given for rapidly decaying particles; for estimations it is convenient to make use of relation (A.1) in the form $\Gamma \, (\mathrm{eV}) \leftrightarrow 10^{-15}/\tau \, (\mathrm{s})$, where the lifetime is expressed in seconds, and the width Γ in electronvolts. In the last columns of the tables particle decay channels are indicated together with the branching ratio of each channel, i.e., the ratio of the decay probability via the given channel to the total decay probability via all possible channels. Thus, from Table A.1 we see that the charged W boson decays, for example, into a positron (electron) and a neutrino (antineutrino), while the neutral Z^0 boson decays into an electron–positron pair e^+e^- or into a $\mu^+\mu^-$ pair.

Table A.2. Properties of leptons ($J = 1/2$)

Particle	Mass m/MeV	Lifetime τ/s	Decay channels
ν_e	$< 17\,\mathrm{eV}$	stable	
ν_μ	< 0.27	stable	
ν_τ	< 35	stable	
e	0.51099906 ± 0.00000015	stable	
μ	105.658387 ± 0.000034	2.197×10^{-6}	$\mu^- \to e^- + \bar{\nu}_e + \nu_\mu$ ($\sim 100\,\%$)
τ	$1784^{+2.7}_{-3.6}$	3×10^{-13}	$\mu\bar{\nu}\nu$ ($\sim 18.5\,\%$) $e\bar{\nu}\nu$ ($\sim 16.5\,\%$) $\pi\nu$ ($\sim 10.3\,\%$) $\rho\nu$ ($\sim 22.1\,\%$) and others

The second group of elementary particles comprises the leptons (Table A.2). This group includes the electron e and the electron neutrino ν_e, the muon μ and the muon neutrino ν_μ, the τ lepton (the "heavy lepton") and the τ neutrino ν_τ. Each of these particles has an antiparticle corresponding to it:

$$
\begin{aligned}
e^- &\to e^+ ; & \nu_e &\to \bar{\nu}_e ; \\
\mu^- &\to \mu^+ ; & \nu_\mu &\to \bar{\nu}_\mu ; \\
\tau^- &\to \tau^+ ; & \nu_\tau &\to \bar{\nu}_\tau .
\end{aligned}
\tag{A.2}
$$

The inherent angular momentum (spin) of each of these particles is $1/2$. Mutual transformations and decays of leptons are regulated by the laws of weak interaction, which include the conservation of energy, momentum, electric charge and angular momentum, as well as the conservation of a quantum number peculiar to leptons, called the lepton charge (to be more precise, conservation of each of the three kinds of lepton charge, L_e, L_μ, and L_τ):

$$
L_e = \begin{cases} +1; e^-, \nu_e , \\ -1; e^+, \bar{\nu}_e , \end{cases} \quad
L_\mu = \begin{cases} +1; \mu^-, \nu_\mu , \\ -1; \mu^+, \bar{\nu}_\mu , \end{cases} \quad
L_\tau = \begin{cases} +1; \tau^-, \nu_\tau , \\ -1; \tau^+, \bar{\nu}_\tau . \end{cases}
\tag{A.3}
$$

We shall take advantage of the muon decay to illustrate the laws of weak interaction. Thus, the channel $\mu^- \to e^- + \nu$, allowed from the point of view of electric charge conservation, is forbidden by spin conservation; both the $\mu^- \to e^- + \nu_e + \nu_e$ and $\mu^- \to e^- + \nu_e + \nu_\mu$ channels are forbidden owing to conservation of the lepton charges L_e and L_μ. The result is that both the μ^+ and the μ^- mesons are each allowed to decay only via a sole decay channel:

$$
\begin{aligned}
\mu^- &\to e^- + \bar{\nu}_e + \nu_\mu , \\
\mu^+ &\to e^+ + \nu_e + \bar{\nu}_\mu ;
\end{aligned}
\tag{A.4}
$$

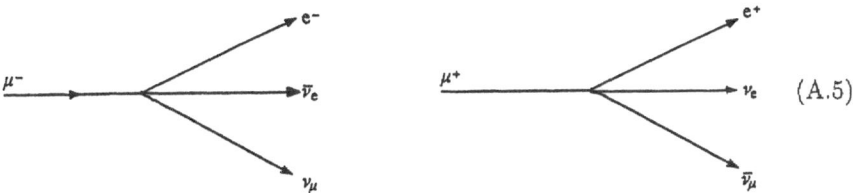

$$\tag{A.5}$$

which imply "point-like" (contact) interaction between leptons. The establishment of the existence of intermediate bosons permitted the revealing of the four-particle weak interaction vertex from the point of view of the exchange mechanism, universal for all types of interaction. Thus, for example, a diagram involving W-boson exchange [see (A.6)] now corresponds to the first diagram of (A.5).

The conservation laws formulated above are satisfied at each of the vertices of this diagram.

In field theory, the mass m of the particle associated with the internal line of a diagram is shown to determine the

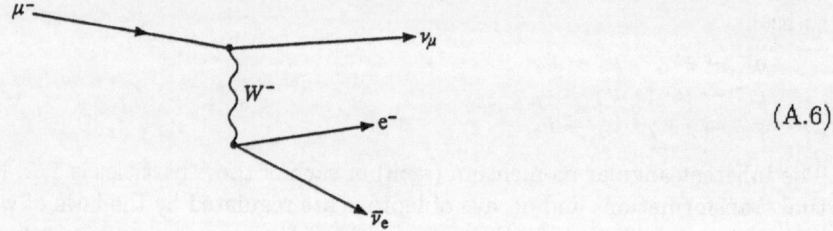

$$\text{(A.6)}$$

"interaction radius", i. e., the space dimensions of the interaction region responsible for the process described by the diagram:

$$a = \frac{\hbar}{mc}. \tag{A.7}$$

Again, for estimations it is convenient to make use of this relation in the following form: a (fm) $\leftrightarrow 200/m$ (MeV), where particle masses are expressed in MeV, and the interaction radius in femtometers ($1\,\text{fm} = 10^{-13}\,\text{cm}$). Since the mass of the intermediate boson exceeds $80\,\text{GeV}$, the weak interaction radius turns out to be of the order of $10^{-15}\,\text{cm}$, which is significantly smaller than typical "nuclear" dimensions of $10^{-13}\,\text{cm}$. Hence, processes of the type in (A.6) are apparently "contact", as compared with strong interaction processes, and, in a certain sense, they can be represented by four-particle interaction diagrams (A.5).

Now we turn to the third, largest, group of elementary particles, the *hadrons* (particles interacting strongly). The set of quantum numbers characterizing their properties reflects a series of specific characteristics. First of all, this is the isospin I (in nuclear physics it is denoted by T), a quantum number assuming either integer, or half-integer values. When two or more hadrons are combined into a single system, their isospins are added to each other according to the addition rules for angular momenta. Thus, the total isospin I_{12} of two hadrons with isospins I_1 and I_2, respectively, assumes a value from the set $I_{12} = |I_1 - I_2|, \ldots, I_1 + I_2$. When $I \neq 0$, a whole group of hadrons called an *isobaric multiplet* is associated with the same isospin value I. They are distinguished from each other by an additional quantum number I_3, that is the projection of the isospin onto the z axis in abstract (isobaric) space and for a given I assumes a discrete set of values

$$I_3 = -I, \ -I+1, \ldots, I-1, \ I; \tag{A.8}$$

the total amount (multiplicity) equals $2I + 1$. Strong interactions exhibit the property of isobaric invariance: this means that hadrons pertaining to one isobaric multiplet always behave identically with respect to strong interactions. However, they have different electric charges and, consequently, behave differently in electromagnetic interactions (which is manifested, for instance, in the differences between the masses of hadrons belonging to the same multiplet). Hadrons of isospin $I = 1$ form isobaric triplets (such, for instance, are the meson triplets π^+, π^0, π^-; ρ^+, ρ^0, ρ^-, and, also, the triplet of Σ hyperons Σ^+, Σ^0, Σ^-; here, $I_3 = +1, 0, -1$, respectively); hadrons of isospin $I = 1/2$ form isobaric doublets [for example, the pair of nucleons: the proton p ($I_3 = +1/2$) and the neutron n ($I_3 = -1/2$)]. In the tables presented below we shall also see isobaric singlets, when $I = 0$ (for example, the η meson or the ω meson), and isobaric quadruplets, when $I = 3/2$ (such, for example, is the resonance $\Delta(1232)$, which is the lowest excited state of a nucleon. It is encountered in four different charge states: Δ^{++}, Δ^+, Δ^0, Δ^-; and $I_3 = 3/2, 1/2, -1/2, -3/2$, respectively).

Another characteristic peculiar to hadrons is *strangeness*. The corresponding quantum number S assumes integer values $S = 0, \pm 1, \dots$. Generalizing the notion of strangeness we arrive at the concepts of *charm* (the corresponding quantum number is denoted by c) and *beauty* (with the quantum number b). All these quantum numbers are conserved in processes for which the strong interaction of hadrons is responsible.

The conservation laws indicated, together with the conservation laws of parity, electric charge, and angular momentum, impose rigorous limits on the character of mutual transformations and decays of hadrons due to strong interaction. The rates of such processes are usually high. Thus, for example, the $\rho(770)$ meson, which is the lightest of the vector mesons ($J^P = 1^-$), has a spread in mass of over 100 MeV (Table A.3); this means that its mean lifetime τ is smaller than 10^{-23} s. Hadrons may participate in the weak interaction. Thus, for instance, precisely weak interaction is repsonsible for hadron decays resulting in the production of leptons. For example:

$$\pi^+ \to \mu^+ + \nu_\mu; \quad \pi^- \to \mu^- + \bar{\nu}_\mu \quad (\tau = 2.6 \times 10^{-8}\,\text{s});$$
$$\text{n} \to \text{p} + \text{e} + \bar{\nu}_\text{e} \quad (\tau \approx 15\,\text{min}). \tag{A.9}$$

The rate of such processes is many orders of magnitude smaller than that of processes due to strong interaction. Parity is not conserved in weak interactions involving hadrons; here, the conservation laws of isospin and strangeness and of other quantum numbers peculiar only to hadrons are also not valid.

Now, consider the data presented in Table A.3. What draws our attention is the large – by a factor of over one hundred – difference between the widths of the η meson and of the next meson closest to it in mass, the ρ meson. Moreover, unlike the ρ meson, the η meson does not decay via the channel $\eta \to 2\pi$, although this decay channel is quite permissible from the point of view of energy. What is the reason of this prohibition? Let us apply to the decay channel of interest,

$$\eta \to 2\pi(?), \tag{A.10}$$

the conservation laws, formulated above, in combination with each other.

The η meson is an isosinglet ($I = 0$); the π mesons form an isotriplet ($I_\pi = 1$). In accordance with the vector rule for adding momenta $1 + 1 = 0, 1, 2$, the decay $\eta(I = 0) \to \pi(I = 1)$ (say, $\eta \to \pi^+\pi^-$) is permitted by isospin conservation. Now, we turn to the parity conservation law. The

Table A.3. Properties of hadrons

Particle	Isospin, spin, parity	m/MeV	$\tau(\Gamma)$	Decay channels
\multicolumn{5}{c}{Nonstrange mesons ($S = 0$)}				
$\pi^{\pm}$	$1(0^-)$	139.57	$2.6 \times 10^{-8}\,\mathrm{s}$	$\mu\nu$ ($\sim 100\%$)
π^0	$1(0^-)$	134.97	$0.84 \times 10^{-16}\,\mathrm{s}$	$\gamma\gamma$ ($\sim 98.8\%$) $\gamma e^+ e^-$ (1.2%)
η	$0(0^-)$	548.8 ±0.6	1.19 ±0.12 keV	$\gamma\gamma$ (39%) $3\pi^0$ (31.8%) $\pi^+\pi^-\pi^0$ (23.7%) and others
$\rho(770)$	$1(1^-)$	768.3 ±0.5	149.1 ±2.9 MeV	$\pi\pi$ ($\sim 100\%$) $\pi\gamma$ (0.05%)
$\omega(783)$	$0(1^-)$	781.9 ±0.14	8.43 ±0.10 MeV	$\pi^+\pi^-\pi^0$ (88.8%) $\pi^0\gamma$ (8.5%)
\multicolumn{5}{c}{Strange mesons ($S = \pm 1$)}				
$K^{\pm}$	$\frac{1}{2}(0^-)$	493.64	$1.237 \times 10^{-8}\,\mathrm{s}$	$\mu\nu$ (63.5%) 2π (21.2%) 3π (7.3%)
$K, \overline{K}^0$	$\frac{1}{2}(0^-)$	497.7	$0.8 \times 10^{-10}\,\mathrm{s}$ (K_s^0)	2π
			$5 \times 10^{-8}\,\mathrm{s}$ (K_l^0)	3π (61%) $\pi\mu\nu$ (39%)
\multicolumn{5}{c}{Nonstrange baryons ($S = 0$)}				
p	$\frac{1}{2}\left(\frac{1}{2}^+\right)$	938.27	$> 10^{32}$ years	
n	$\frac{1}{2}\left(\frac{1}{2}^+\right)$	939.57	$\sim 15\,\mathrm{min}$	$p\,e\,\bar{\nu}_e$
$\Delta(1232)$	$\frac{3}{2}\left(\frac{3}{2}^+\right)$	1232	$\sim 115\,\mathrm{MeV}$	$N\pi$
$N(1440)$	$\frac{1}{2}\left(\frac{1}{2}^+\right)$	1440	$\sim 200\,\mathrm{MeV}$	$N\pi$ ($\sim 55\%$) and others

Table A.3 (continued)

Particle	Isospin, spin, parity	m/MeV	$\tau(\Gamma)$	Decay channels
Strange baryons (hyperons) $(S \neq 0)$				
Λ	$0\left(\frac{1}{2}^+\right)$ $S = -1$	1115.6	2.6×10^{-10} s	$p\pi^-$ (64.2%) $n\pi^0$ (35.8%)
Σ^+	$1\left(\frac{1}{2}^+\right)$ $S = -2$	1189.37	0.8×10^{-10} s	$N\pi$
Σ^0	$1\left(\frac{1}{2}^+\right)$ $S = -2$	1192.55	7.4×10^{-20} s	$\Lambda\gamma$
Σ^-	$1\left(\frac{1}{2}^+\right)$ $S = -2$	1197.43	1.48×10^{-10} s	$N\pi$
Ξ^0	$\frac{1}{2}\left(\frac{1}{2}^+\right)$ $S = -2$	1314.9	2.90×10^{-10} s	$\Lambda\pi$
Ξ^-	$\frac{1}{2}\left(\frac{1}{2}^+\right)$ $S = -2$	1321.32	1.64×10^{-10} s	$\Lambda\pi$
Hadrons with charm (c) and beauty (b)				
$D^\pm$	$\frac{1}{2}\left(0^-\right)$ $c = \pm1$	1869.3 ±0.4	1.06×10^{-12} s	$K + \ldots$ ($\sim 70\%$)
Λ_c^+	$0\left(\frac{1}{2}^+\right)$ $c = 1$	2285.2 ±1.2	1.91×10^{-13} s	$\Lambda(\Sigma) + \ldots$ ($\sim 50\%$)
$B^\pm$	$\frac{1}{2}\left(0^-\right)$ $b = \pm1$	5277.6 ±1.4	1.18×10^{-12} s	$K + \ldots$ ($\sim 80\%$)

intrinsic parity of the η meson is negative ($P_\eta = -1$); the same is true of the intrinsic parity of pions ($P_\pi = -1$). In strong interactions, parity is conserved. This means that the total parity of the 2π system in the final state of process (A.10) must be negative. On the other hand, it is the product of the intrinsic parities of two pions and of the parity of their relative motion in the 2π system, which, in turn, is determined by the angular momentum L of this relative motion. Thus,

$$P_{2\pi} = P_\pi P_\pi (-1)^L = (-1)^L . \tag{A.11}$$

For this quantity to be negative (and, consequently, for it to coincide with the intrinsic parity of the η meson, $P_\eta = -1$), the angular momentum L of the system composed of two pions produced in process (A.10) must be odd, i.e., nonzero, but since the pion is a spinless particle ($J^P = 0^-$), the total angular momentum of the 2π system must also be nonzero in this process. At the same time, the total angular momentum of the system considered equals zero in the initial state of the process (the η meson is a spinless particle, also: $J^P = 0^-$). Thus, the decay (A.10) turns out to be strictly forbidden by the combination of the conservation laws of angular momentum and parity in strong interactions. It can be shown that in the case of the three-pion decay channel $\eta \to 3\pi$ the same conservation laws do not lead to prohibition; only, in this case a considerably more complex rearrangement of the internal hadron structure is required, so the $\eta \to 3\pi$ decay is much slower than, say, the decay $\rho \to 2\pi$.

Below we present examples of diagrams for hadron decay processes. The decay of the charged pion (weak interaction) is:

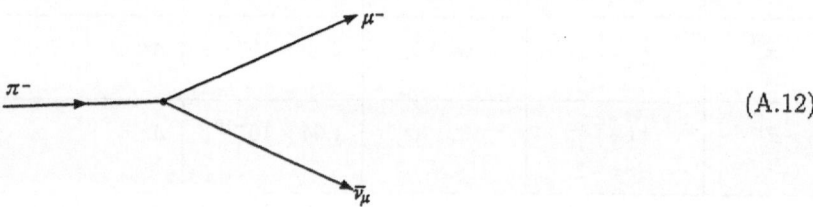

(A.12)

The decay of the neutral pion (electromagnetic interaction) is:

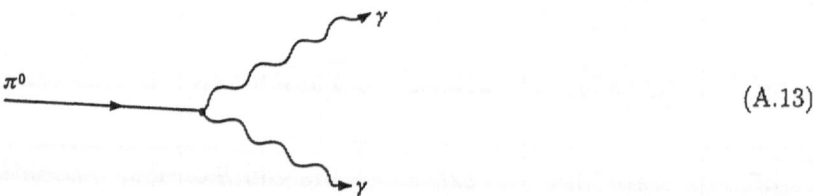

(A.13)

The $\pi^0 \to 2\gamma$ is the main channel for the π^0-meson decay. From Table A.3 one can also see that the $\pi^0 \to \gamma e^+ e^-$ channel also contributes about one percent to the total probability of the neutral pion decay. The corresponding diagram follows:

(A.14)

Here, the internal line corresponds to a virtual photon, whereas the lower vertex reflects its transformation into the electron–positron pair. The ratio of the decay probabilities via the respective channels (A.14) and (A.13) can be estimated by taking into account that the decay amplitude for (A.14) $A(\pi^0 \to \gamma e^+ e^-)$, as compared with the amplitude for $A(\pi^0 \to 2\gamma)$ involves an extra factor – the electromagnetic interaction constant e (the electron charge):

$$A(\pi^0 \to \gamma e^+ e^-) \sim \frac{e}{\sqrt{\hbar c}} A(\pi^0 \to 2\gamma) \qquad (A.15)$$

(the factor $1/\sqrt{\hbar c}$ makes the expression $e/\sqrt{\hbar c}$ dimensionless). Hence, for the ratio of the π_0-meson decay probabilities via the two respective channels we obtain the estimate

$$\frac{W(\pi^0 \to \gamma e^+ e^-)}{W(\pi^0 \to 2\gamma)} = \left| \frac{A(\pi^0 \to \gamma e^+ e^-)}{A(\pi^0 \to 2\gamma)} \right|^2 \sim \frac{e^2}{\hbar c} = \frac{1}{137}, \qquad (A.16)$$

which, as one can see from Table A.3, is in qualitative agreement with experimental data.

Table A.4. Characteristics of quarks

Sort of quark (flavor)	Charge	Mass	Isospin	S	C	b	t
u	2/3	1 MeV	1/2	0	0	0	0
d	−1/3	1 MeV	1/2	0	0	0	0
s	−1/3	150 MeV	0	−1	0	0	0
c	2/3	2 GeV	0	0	1	0	0
b	−1/3	5 GeV	0	0	0	−1	0
t	2/3	170 GeV	0	0	0	0	1

In modern theory of elementary particles leptons and intermediate bosons are considered to be structureless particles, whereas hadrons consist of *quarks*. The theory requires six sorts of quarks. All of them have been discovered experimentally as constituents of hadrons (Table A.4). They have fractional charge and, most likely, cannot exist in a free state.

From a quantitative point of view the quark theory of elementary particles (*quantum chromodynamics*, QCD) is very complicated. This is due to the special nature of the exchange interaction between quarks. At first sight, the interaction between two quarks q_1 and q_2 seems similar to the interaction between two electric charges (say, between two electrons) in quantum electrodynamics (QED):

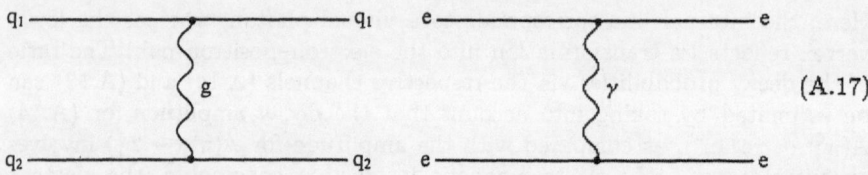

$$(A.17)$$

The interaction of charges is realized owing to the exchange of a virtual photon; in quantum chromodynamics this role is assumed by the gluon g. However, gluons, unlike photons, are capable of interacting directly with each other. In QCD, for instance, the following simplest diagram is possible:

$$(A.18)$$

while in QED the interaction between photons is described by only the 4th order diagram

$$(A.19)$$

(the probability of such interaction is proportional to $(e^2/\hbar c)^4$ and is, consequently, very small).

Owing to processes such as (A.18), the interaction between quarks turns out to be totally unlike anything encountered in elementary particle physics of the "prequark" period. Here, enhancement (instead of weakening, as usual) of the interaction of two particles (quarks) with each other takes place as the distance between the particles increases, which results in the phenomenon of quark "confinement" inside the hadrons, so we know nothing of quarks in a free state. Quark theory can be seen to develop along two lines: consistent, rigorous, but extremely cumbersome (possible only if the most powerful computers are utilized) calculations within QCD, and models allowing revelation or stressing of some aspect of the hadron quark structure. These two lines of investigation are closely related.

Tables A.5 and A.6 present the quark compositions of hadrons. Baryons consist of three quarks qqq, mesons of a quark q and an antiquark q̄ or, in certain cases, of combinations of (qq̄) pairs. Note the quark structure of the η meson. Unlike the π meson, which contains no quarks or antiquarks with strangeness, and, on the other hand, unlike K mesons, which are strange

Table **A.5.** Quark compositions
of some baryons

Baryon	Quark composition
p	uud
n	udd
Λ	uds
Δ^{++} (1232)	uuu
Ξ^0	uss

Table **A.6.** Quark compositions
of some mesons

Meson	Quark composition
π^+	$u\bar{d}$
π^0	$u\bar{u}$, $d\bar{d}$
K^+	$u\bar{s}$
η	$u\bar{u}$, $d\bar{d}$, $s\bar{s}$
D^-	$\bar{c}d$

hadrons ($S = \pm 1$), the η meson is not a strange hadron, although the strange
quark s and the antiquark $\bar{s}$ are among its constituents (it would be more
correct to say that the η meson passes some time in the form of an $s\bar{s}$ system).
In this case we speak of a hadron with *hidden strangeness*. A similar concept
is that of *hidden charm*; both strange and nonstrange heavy mesons, which
contain the $c\bar{c}$ pair, have hidden charm.

Classification of elementary particles is based on the group-theoretical
approach. The three quarks u, d, s make up the basis for the fundamental
representation of the unitary group SU_3. Applying the general rules of group
theory, one can group the hadrons of identical baryon charge B and identical
spin and parity J^P (such hadrons have similar, nearly coinciding, intrinsic
space structures), but differing in electric charge, isospin, and strangeness into
certain SU_3 multiplets: octets [8], decouplets [10], and so on; some hadrons
do not group with others and are SU_3 singlets [1]. Figures A.1–A.3 present
examples of baryon and meson octets and of a decouplet and illustrate the
quark composition of the hadrons pertaining to the given multiplet. The
isospin projections I_3 and the *hypercharge* $Y = B + S$, a quantum number
determined by the baryon charge B of a particle and by its strangeness S,
are indicated along the horizontal and vertical axes, respectively.

In the language of quarks, the diagrams for various processes involving the
participation of hadrons can be depicted in a different way. Thus, for instance,
the decay of the Δ isobar, $\Delta \rightarrow N + \pi$, is interpreted as the production
of a quark–antiquark pair involving rearrangement of the quarks inside the
baryon:

$$(A.20)$$

In a similar way the structure of the $NN\pi$ vertices is revealed in the strong-
interaction diagrams:

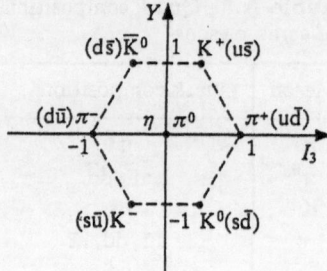

Fig. A.1. Octet of pseudoscalar mesons $(J^P = 0^-)$

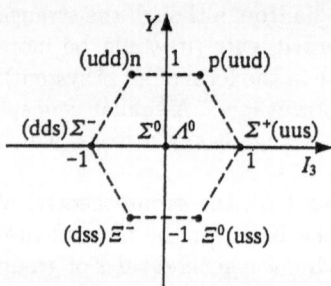

Fig. A.2. Baryon octet $(J^P = 1/2^+)$

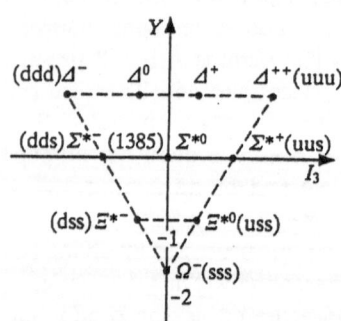

Fig. A.3. Baryon decouplet $(J^P = 3/2^+)$

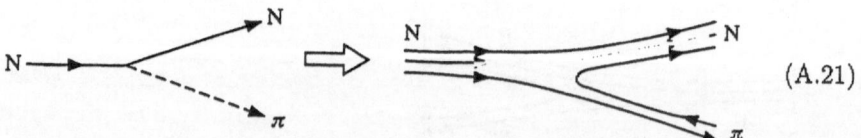

$$(A.21)$$

As to the weak interaction processes, in which hadrons participate, the diagrams of all such processes can be depicted in the quark language in a uniform manner with the aid of the elementary diagram for the interaction of a quark (antiquark) with an intermediate boson:

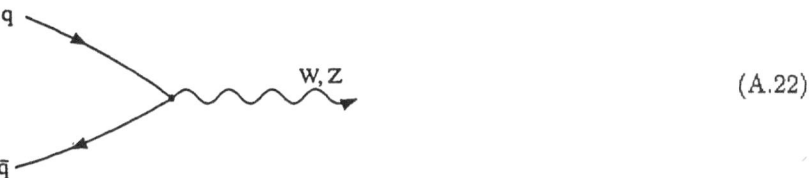

(A.22)

Such, for instance, is the quark diagram representing the pion decay $\pi \to \mu + \nu_\mu$:

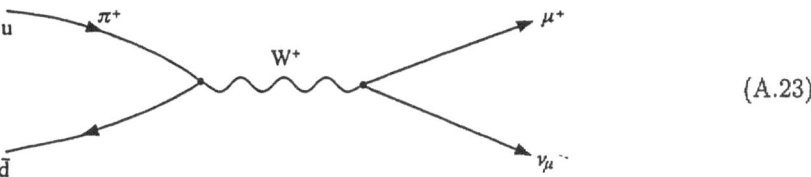

(A.23)

Another example is the decay of the Λ hyperon, $\Lambda \to p + \pi^-$:

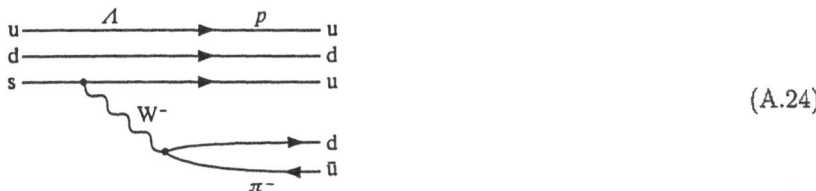

(A.24)

Here, the conservation of strangeness is violated at the suW vertex.

Turning to the issue of quark masses, one must distinguish between the masses of "bare" quarks (the corresponding values are given in Table A.4, and precisely these values are involved as the "initial", "bare" quark masses in consistent QCD calculations) and the masses of quarks serving as constituent particles of hadrons ("constituent" quarks), when the quark mass includes the entire energy of the gluon field per single constituent quark. From the mass of a nucleon (~ 1 GeV), it follows that the mass of a constituent u or d quark amounts to approximately 300 MeV; the mass of a constituent s quark turns out, on the basis of the Λ-hyperon mass, to be about 150 MeV higher (such is the difference between the "bare" masses m_s and m_{ud}).

One of the most simple, but extremely useful, hadron models is the quark bag model (Fig. A.4). According to this model, quark confinement in hadrons is due to certain forces acting from outside, from the "vacuum", on the quark bag, which does not permit it to expand indefinitely. The quarks inside the bag are considered to be free, they do not interact with each other. In its simplest version, the quark bag model has a sole (universal for all hadron systems) parameter: the pressure P_{vac} of vacuum on the bag. The total energy of a hadron consisting of N quarks is the sum of the total energy of the free

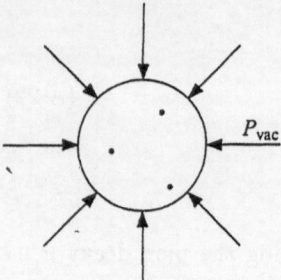

Fig. A.4. Quark bag model

quarks $\Sigma\varepsilon_q$ "confined" inside the bag and the volume energy $\frac{4\pi}{3}R^3 P_{vac}$ (where R is the radius of the "bag") related to the work performed by the external pressure P_{vac} squeezing the bag. The first term decreases as R increases. We shall consider the nucleon, for definiteness. The masses of the u and d quarks ($\sim$1 MeV) are negligible compared with their total energy ($\sim$300 MeV), i. e., the quarks in the bag are ultrarelativistic. For particles with $m_q \ll \varepsilon_q$ the relation between energy and momentum, $\varepsilon_q = \sqrt{p^2c^2 + m^2c^4}$, transforms into $\varepsilon_q = p_q c$, where p is the momentum. On the other hand, owing to the uncertainty relation for the average momentum of a particle inside a bag of radius R we have $p_q \approx h/R$. Hence it follows that the energy of a quark in a bag of radius R is a quantity of the order of $\varepsilon_q \sim hc/R$. The exact expression for the spectrum of energy levels of a massless particle in a well can be derived from the Dirac equation. Thus, for example, the following expression is valid for the ground state:

$$\varepsilon_q = 2.04 \frac{\hbar c}{R}\,. \tag{A.25}$$

Thus, quarks moving freely tend to occupy the largest possible volume, but the pressure exerted by the vacuum on the bag impedes their flying apart. The equilibrium dimensions of a bag of R_0 are determined by the condition

$$\partial E_{tot}/\partial R = 0\,, \tag{A.26}$$

where

$$E_{tot}(R) \approx 2N\frac{\hbar c}{R} + \frac{4\pi}{3}R^3 P_{vac} \tag{A.27}$$

is the total energy (mass) of a hadron composed of N quarks (Fig. A.5). For a nucleon ($N = 3$) we hence have

$$m_N c^2 \approx \frac{8\hbar c}{R_0}\,; \quad R_0 \approx \left(\frac{3\hbar c}{2\pi P_{vac}}\right)^{1/4}\,. \tag{A.28}$$

Choosing $P_{vac} \approx 20$ MeV/fm^3, we obtain

$$m_N \approx 1\,\text{GeV}\,; \quad R_0 \approx 1.6\,\text{fm}\,,$$

which is in quite good agreement with the actual values of the mean nucleon dimensions and mass.

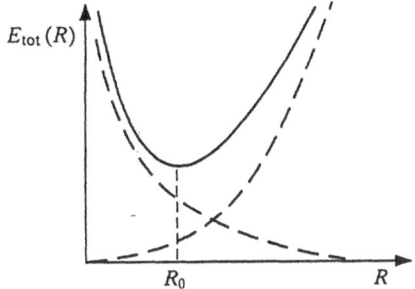

Fig. A.5. Relation between the kinetic energy of the quarks and the volume energy in the quark bag

B. The Atomic Nucleus

The physics of the atomic nucleus is closely related to elementary particle physics by the problem of nuclear forces. The carriers of strong interaction between nucleons are mesons:

$$
\begin{array}{c}
N \longrightarrow \longrightarrow N \\
\big| \\
\big|\; \pi, \rho, \omega, \ldots \\
\big| \\
N \longrightarrow \longrightarrow N
\end{array}
\qquad (B.1)
$$

The lightest meson, the pion, is responsible for the attractive part of atomic interaction at large (on a nuclear scale) distances $\gtrsim 1\,\mathrm{fm}$. As the nucleons approach each other, the exchange of other heavier mesons becomes more and more significant, and the general character of the interaction is essentially altered: the attraction transforms into repulsion, which at very small distances ($\lesssim 0.4\,\mathrm{fm}$) acquires the form of a strong repulsive "core". At present, meson theory, which is based on knowledge of the structure of elementary particles and on quantum field theory, is the main instrument for qualitative and quantitative descriptions of nuclear forces. The quark approach to this problem is also developing rapidly.

Let us consider the simplest meson theory of nuclear forces, classical mesodynamics. A static meson field $\varphi(\boldsymbol{r})$ produced by sources with a density distribution $\rho(\boldsymbol{r})$ satisfies the equation

$$
\left[\Delta - \left(\frac{mc}{\hbar}\right)^2\right]\varphi(\boldsymbol{r}) = -4\pi g\,\rho(\boldsymbol{r})\,,
\qquad (B.2)
$$

where m is the mass of the quantum (meson) of the field considered. This equation is a generalization of the Poisson equation in electrodynamics,

$$
\Delta\varphi_{\mathrm{e}}(\boldsymbol{r}) = -4\pi e\rho_{\mathrm{e}}(\boldsymbol{r})\,;
$$

instead of the electric charge e, which is the electromagnetic constant, this equation contains the strong interaction coupling constant g. A more correct form of (B.2) and, consequently, the properties of its solutions depend on the transformation properties of the field $\varphi(r)$ with respect to rotations of the space coordinate system, to inversion (reflection) of the coordinates, and to rotations in isospin space. These transformation properties of the meson field, in turn, depend on the spin, internal parity, and isospin of the meson.

The pion is a particle of spin $J^P = 0^-$ and isospin $T = 1$. The corresponding field φ_π is pseudoscalar with respect to transformations in ordinary space and three-dimensional from the point of view of its isospin properties (isovector). When substituting such a function $\varphi_\pi(r)$ into (B.2), it is also necessary to substitute into its right-hand part an expression that is a pseuedoscalar in ordinary space and a vector in isospin space. We shall do so for the nucleon, neglecting its intrinsic size, i. e., considering it to be a point-like source of the pion field:

$$\left[\Delta - \left(\frac{m_\pi c}{\hbar}\right)^2\right]\varphi(r) = -4\pi f\hat{\tau}(\sigma\nabla)\delta(r), \tag{B.3}$$

where σ is a vector composed of the Pauli spin matrices $\hat{\sigma}_x$, $\hat{\sigma}_y$, $\hat{\sigma}_z$, and $\hat{\tau}$ is a similar vector in isospin space; f is the strong interaction constant. When the source of the pion field is a particle of spin $1/2$, the expression presented is the only one exhibiting a structure that satisfies the necessary transformational requirements.

Equations such as (B.2) are conveniently solved with the aid of the Fourier transformation:

$$\varphi(r) = \int \varphi_k e^{ikr} d^3k. \tag{B.4}$$

Applying this method in the case of (B.3), we obtain for our pion field

$$\varphi_\pi(r) = \frac{f}{m_\pi c/\hbar}\hat{\tau}(\sigma\nabla)\frac{1}{r}\exp\left(-\frac{m_\pi c}{\hbar}r\right). \tag{B.5}$$

The average range $r_0 = \hbar/m_\pi c$ of this field is determined by the pion mass [see general expression (A.7)]. The energy of a second nucleon placed in this field is given by the expression

$$V(r) = f\tau_2(\sigma_2\nabla)\varphi_\pi^{(1)}(r), \tag{B.6}$$

where the index (1) signifies that the operators $\hat{\tau}$ and σ, present in $\varphi_\pi^{(1)}(r)$, are related to the first nucleon: $\hat{\tau} \to \hat{\tau}_1$, $\sigma \to \sigma_1$; the interaction constant f is the same one as in (B.5) (once again, it is possible to verify that only expression (B.6) has a structure permitting one to use the pseudoscalar function $\varphi_\pi(r)$ and the nucleon operators $\hat{\tau}$ and σ for constructing invariants relative to all the above indicated transformations of the coordinate system). Substituting (B.5) into (B.6), we finally obtain for the interaction energy of two nucleons

$$\hat{V}(r_{12}) = \frac{1}{3}\frac{f^2}{\hbar c}m_\pi c^2(\hat{\tau}_1 \cdot \hat{\tau}_2)\left[\sigma_1\sigma_2 + \left(1 + \frac{3}{\mu r} + \frac{3}{(\mu r)^2}\right)\hat{S}_{12}\right]\frac{e^{-\mu r}}{\mu r}, \tag{B.7}$$

where, for brevity, we have introduced the notation $\mu = m_\pi c/\hbar$. Here, $\hat{S}_{12}$ is the interaction tensor operator of the two nucleons:

$$\hat{S}_{12} = \frac{3(\sigma_1 r)(\sigma_2 r)}{r^2} - (\sigma_1 \sigma_2). \tag{B.8}$$

The eigenvalues of the spin operators $(\sigma_1 \sigma_2)$ and $\hat{S}_{12}$ are expressed via the total spin of the two nucleons $S = 1/2(\sigma_1 + \sigma_2)$ and (in the case of the operator $\tilde{S}_{12}$) its projection $S_z = (Sr)/r$ onto the vector $r = r_1 - r_2$ reaching from one of these nucleons to the other:

$$\langle \sigma_1 \sigma_2 \rangle_{SS_z} = 2S(S+1) - 3 \,,$$
$$\langle S_{12} \rangle_{SS_z} = 6S_z^2 - 2S(S+1) \,. \tag{B.9}$$

In the same manner, the mean scalar product $(\hat{\tau}_1 \cdot \hat{\tau}_2)$ is expressed via the total isospin T of the nucleon pair under consideration:

$$\langle \hat{\tau}_1 \cdot \hat{\tau}_2 \rangle_{TT_z} = 2T(T+1) - 3 \,. \tag{B.10}$$

At low energies, interaction in the s state plays a most important role. According to the Pauli principle, the nucleon pair, in this case, may be in one of the two following spin–isospin states: $|S = 0,\ T = 1\rangle$ or $|S = 1,\ T = 0\rangle$. In the first of these states the potential (B.7) represents the Yukawa attraction potential

$$V(r_{12})\Big|_{S=0,T=1} = -\frac{f^2}{\hbar c} m_\pi c^2 \frac{e^{-\mu r}}{\mu r} \,. \tag{B.11}$$

In the states with $S = 1$ additional attraction occurs owing to tensor forces. Thus, the exchange of a pseudoscalar isovector meson (the pion) not only explains the very existence of strong interaction between nucleons, but also results in the characteristic spin dependence of nuclear forces. The amplitude of the interaction potential depends on the coupling constant f of the pion–nucleon interaction. According to the data on pion–nucleon scattering it can be found from the relation

$$\frac{f^2}{\hbar c} \approx 0.08 \,. \tag{B.12}$$

This dimensionless parameter of strong-interaction theory is significantly greater than the corresponding parameter $e^2/\hbar c = 1/137$ present in the theory of electromagnetic interactions. But, anyhow, it is smaller than unity. Therefore, the exchange of several pions in the nucleon interaction potential (at distances $r \geq 1.5$–$2\,\text{fm}$) is an effect of a higher order of magnitude and results in no qualitative change in the properties of this interaction, which are due to one-pion exchange.

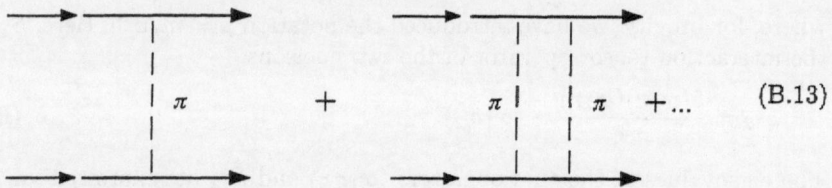

$$(B.13)$$

The idea of the dominant contribution of one-pion exchange underlies the so-called OBEP (one-boson exchange potential) approach. Potentials constructed within the framework of this approach pertain to the class of semiphenomenological potentials: their shape is determined by the general rules for constructing diagrams like (B.1), whereas for the choice of the optimum number of mesons taking part in the exchange, the numerical values of the coupling constants, the repulsion parameters at small distances, and so on, the last word here is due to the procedure for fitting the experimental data. The general shape of the NN interaction potential is presented schematically in Fig. B.1.

The main features peculiar to nucleon–nucleon interaction are clearly manifested in the properties of the simplest nucleus, the deuteron. The deuteron has a sole bound state with quantum numbers $J^\pi = 1^+$, $T = 0$ (the binding energy $\varepsilon_D = 2.23\,\text{MeV}$). This state, which is a spin triplet ($S = 1$), is not a pure state relative to the angular momentum of the pair: the tensor interaction between the proton and the neutron leads to the main S component of the deuteron wave function having a small D-wave contribution added to it:

$$\varphi_d = \sqrt{1 - |\alpha_D|^2}\,|np : {}^3S_1\rangle + \alpha_D|np : {}^3D_1\rangle. \qquad (B.14)$$

The D-wave admixture causes the deuteron to exhibit an electric quadrupole moment ($Q_d = 2.7 \times 10^{-27}\,\text{cm}^2$), which definitely points to a deviation from spherical symmetry in the nuclear structure; the D-wave admixture also affects the magnetic moment of the deuteron. Estimates of the D-wave admixture in the deuteron vary within 5–8 %. In spite of this value seeming to be quite small, the admixture plays an extremely important role in many processes involving the deuteron (Fig. B.2).

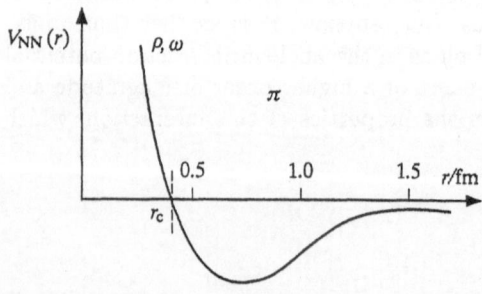

Fig. B.1. Schematic NN interaction potential (the symbols in the upper part of the picture indicate the ranges, where one-pion exchange and the exchange of the vector ρ and ω mesons are dominant; r_c is the radius of the repulsive core)

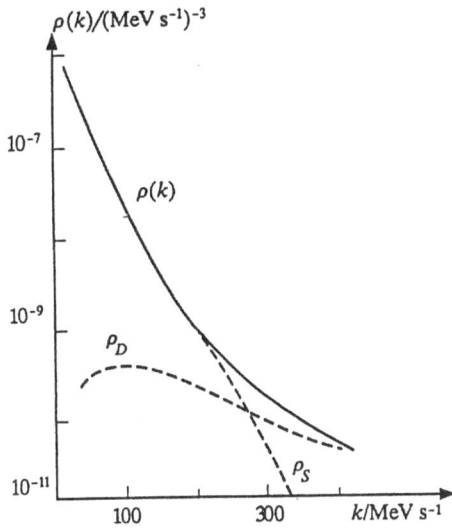

Fig. **B.2.** Momentum distribution of the nucleons in the deuteron $\rho(k) = \rho_S(k) + \rho_D(k)$ showing the contributions of the S and D waves

The interaction between the proton and the neutron in the singlet state $(S = 0)$ is not sufficiently strong to give rise to a bound state in the deuteron. At the same time there also exists no bound state in the neutron–neutron and proton–proton systems, which together with the neutron–proton system make up an isotopic triplet $(T = 1)$. Available data on scattering parameters of the np, nn, and pp systems permit us to draw a conclusion on the accuracy with which the charge independence of nuclear forces is realized. Below, we present the scattering lengths a_{NN} and the effective radii r_{NN} (all in units of 10^{-13} cm) for these systems. For the $^1S_0$ state

$$a_{np} = -23.75, \quad a_{nn} = -18.45, \quad a_{pp} = -17.9,$$

$$r_{np} = 2.75, \quad r_{nn} = 2.83, \quad r_{pp} = 2.82$$

(the values presented for a_{pp} and r_{pp} do not take into account the influence of the electromagnetic interaction between the protons). In the case of the $^3S_1$ state

$$a_{np} = 5.424 \pm 0.004,$$

$$r_{np} = 1.76 \pm 0.005.$$

We shall now consider the structure of compound nuclei. To explain features peculiar to this structure on the basis of nuclear forces and to express all the characteristics of nuclei via the NN interaction – this was the principal goal of the proton–neutron theory of the atomic nucleus formulated from its very origin in the 1930s. Today, such a formulation of the problem seems quite oversimplified. First of all, even if one considers the NN interaction to be known sufficiently well, modern many-body quantum theory is not even capable of calculating, on such a basis, the most simple characteristics of compound nuclei (besides the deuteron, only the three-particle nuclei ^{3}H and ^{3}He permit such a consistent approach). On the other hand, modern studies in nuclear physics reveal with ever greater clarity that the NN interaction in nuclear matter differs essentially from the interaction between two free nucleons. Therefore, nuclear models still play an important part in modern nuclear physics, just like at the earlier stages of its development.

A nonspecialist is always impressed by the enormous number of such models, some of which at first sight even seem to contradict each other: the drop model, the rotating ellipsoid model, the shell model, the cluster, the superfluid, the optical models, and so on. Actually, there is no contradiction between them; each model permits the consideration of nuclei from its own standpoint.

Within the optical model the scattering of a proton or of a neutron by a nucleus is described with the aid of a complex optical potential:

$$V_{\text{opt}}(r) = U_0(r) + iW_0(r) + [U_{sl}(r) + iW_{sl}(r)](\boldsymbol{\sigma l}), \qquad (B.15)$$

the imaginary part of which corresponds to absorption of the nucleon in the nuclear matter. Owing to spin–orbital interaction, the amplitude of the optical potential depends on how the spin and angular momentum of the nucleus add up to yield the total momentum

$$\langle \boldsymbol{\sigma l}\rangle_j = \begin{cases} -(l+1), & j = l + 1/2; \\ l, & j = l - 1/2. \end{cases} \qquad (B.16)$$

The radial dependence of the real part of the central potential $U_0(r)$ is usually given by the Woods–Saxon distribution

$$U_0(r) = -V_0 \frac{1}{1 + \exp((r - R)/a)}, \qquad (B.17)$$

where $R = r_0 A^{1/3}$ is the radius of the nucleus, a is the thickness of its surface layer in which the potential undergoes the most drastic change (Fig. B.3). The shape of the imaginary part of the potential varies, essentially depending on the energy of the incident nucleon: at low energies absorption takes

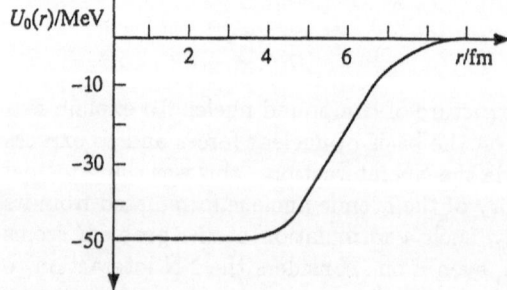

Fig. B.3. Woods–Saxon distribution (B.17) for the real part of the optical interaction potential between nucleons and a nucleus ($A = 100$)

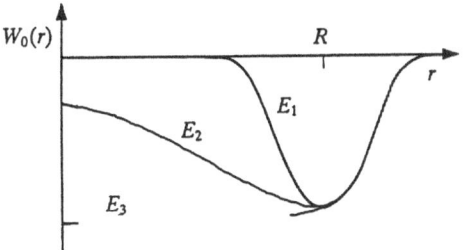

Fig. B.4. Shape of the imaginary part $W_0(r)$ of the optical potential versus the nucleon energy: $E_1 < E_2 < E_3$

place mainly near the surface of the nucleus, whereas as the nucleon energy rises, absorption taking place inside the volume becomes more pronounced (Fig. B.4).

The real part of the optical potential (B.15) is quite well approximated by the mean field potential. Not only the nucleon impinging upon the nucleus, but also each of the nucleons present inside the nucleus "feels" this mean field. The mean field approximation underlies a whole series of models applied for describing the structure of the nucleus. The simplest of these models is the one-particle shell model, which totally neglects all the correlations (residual pair interactions) between the nucleons of the nucleus. The main achievement of this model is the explanation of the "magic" numbers corresponding to the most stable nuclei, in which occupation of the proton and neutron shells is completed (Fig. B.5). For a long time, only indirect, even though quite diverse, experimental facts lay behind the concept of the shell

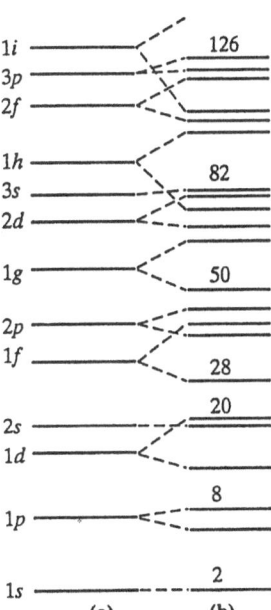

Fig. B.5a, b. Scheme of one-particle nucleon levels in nuclei: (a) without account, (b) with account of spin–orbit splitting

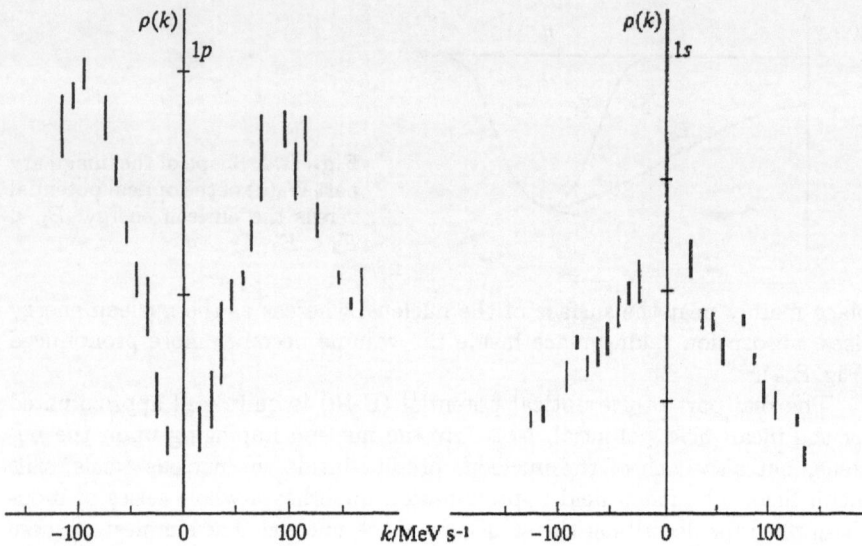

Fig. B.6. Momentum distribution of protons on $1p$ and $1s$ shells in ^{12}C nucleus: measurements of quasielastic proton knockout by electrons by applying the $(e, e'p)$ method (see [B.1])

structure of nuclei. Decisive direct arguments were obtained by the end of the 1950s, when the first experiments were performed in which the quasielastic knockout of protons from nuclei by fast protons was observed in the (p, 2p) reaction. The method of coincidences applied in studies of (p, 2p), (p, np), (e, e'p) reactions made it possible not only to actually "see" from which shell in the nucleus a particular proton was knocked out, but also to feel its wave function on one or another of the shells in the nucleus, and to determine the shape of its momentum distribution inside the nucleus (Fig. B.6). At the same time, effects of residual interactions between the nucleons in the nucleus were quite clearly manifested in quasielastic knockout reactions. The quasielastic knockout of a nucleon from some shell leads to the resultant "hole" interacting with its nucleonic surrounding and "spreading out" over a large number of states of the daughter nucleus, and this spread may be of the order of magnitude of the distance between the shells themselves (Fig. B.7).

For an explanation of the spectrum of excited states of a nucleus it is absolutely necessary to take into account the residual interaction between the nucleons. Such calculations are performed, for instance, in various versions of the many-body model. The one-particle nucleon wave functions, the forms of which depend on the nature of the mean field, play the role of a basis in such calculations aimed at constructing multiparticle wave functions of the nucleus. If the residual interaction is chosen appropriately (so as to include pairing forces and long-range correlations between the nucleons), this approach permits us to obtain, on a sole basis, excited states of differing

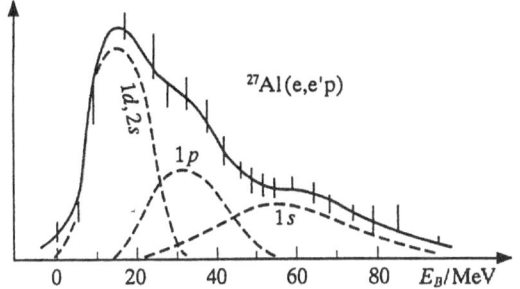

Fig. B.7. Spread of hole excitations in ^{27}Al nucleus: measurements by (e, e′p) method (see [B.2])

nature, both one-particle and collective. In the latter case, the wave function of the excited nucleus constructed within the framework of the many-body model represents a superposition of a large number of elementary excitations.

Figure B.8 presents the scheme of energy levels in the ^{172}Yb nucleus. The dashes show the results of calculations performed within the framework of the so-called IBM model (interacting boson model). Without going into details, we note its "microscopic" character: the one-particle wave functions of the individual nucleons serve as the starting point for describing the nuclear states; they underlie the construction of "elementary" excitations of the nucleus (bosons), which are not, however, independent, but interact with each other. Assumptions concerning the deformation of the nucleus and violation of the mean field symmetry are not taken into account at the beginning. But, in spite of this, the final results of the calculations clearly reveal the presence of a large static and dynamic deformation of the nucleus. This, first of all, is the quite pronounced rotational band of levels 2^+, 4^+, ..., 12^+, based on the ground state 0^+; its levels are positioned in accordance with the rotation law of a strongly deformed nucleus

$$E_J = \frac{\hbar^2}{2I} J(J+1), \tag{B.18}$$

where I is the nuclear moment of inertia. Moreover, calculations permit the identification of vibrational level bands in the excitation spectrum of the nu-

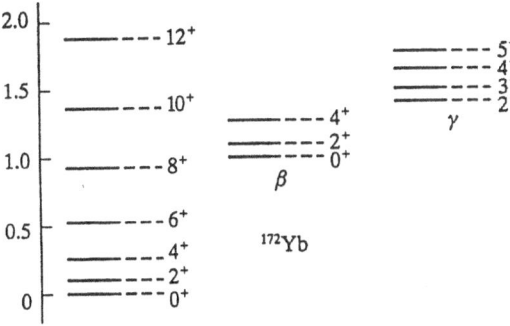

Fig. B.8. The main rotational, β vibrational, and γ vibrational bands of levels in ^{172}Yb nucleus (dashes indicate calculations within IBM model) [B.3]

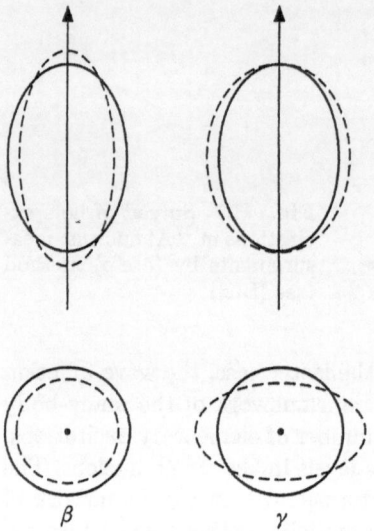

Fig. B.9. Character of β and γ oscillations of a spheroidal nucleus. *Top:* the projection of the nucleus shape onto the plane parallel to the symmetry axis. *Bottom:* the projection onto the plane perpendicular to the symmetry axis of the nucleus

cleus considered. One of these bands $(0^+, 2^+, 4^+)$ corresponds to so-called β oscillations, and the other one $(2^+, 3^+, 4^+)$ corresponds to γ oscillations of the spheroidal nucleus. In the first case the nucleus retains the shape of a rotation ellipsoid, and in the second the axial symmetry of the nucleus is violated (Fig. B.9). Knowledge of the wave functions of nuclear states permits us to calculate the probabilities of electromagnetic transitions between the states. Both calculations and experimental data reveal that transitions between levels of the same band are especially strong – this is a manifestation of the collective character of the corresponding nuclear excitations.

In the example being examined the rotational band is relatively small: from Fig. B.8 it can be seen to terminate at level 12^+. More recent experiments provide examples of very long rotational bands, when the rotational momentum of the nucleus in highly excited states may exceed several tens of units. At present, the investigation of nuclear states of high momenta is a developing branch of nuclear physics.

On the whole, in evaluating the achievements and general possibilities presented by the "microscopic" (i. e., based on nucleonic degrees of freedom) description of nuclear properties, we can formulate the main point: modern "microscopic" models are successful in dealing, on a sole basis, with the interpretation of various collective and one-particle nuclear excitations. At earlier stages in the development of nuclear theory, various models were constructed and used for their explanation (the liquid-drop model explained vibrational excitations, the symmetric or asymmetric ellipsoid model was for rotational excitations, and so on). The above theory is totally relevant to the special class of collective excitations in nuclei: the giant resonances (Fig. B.10).

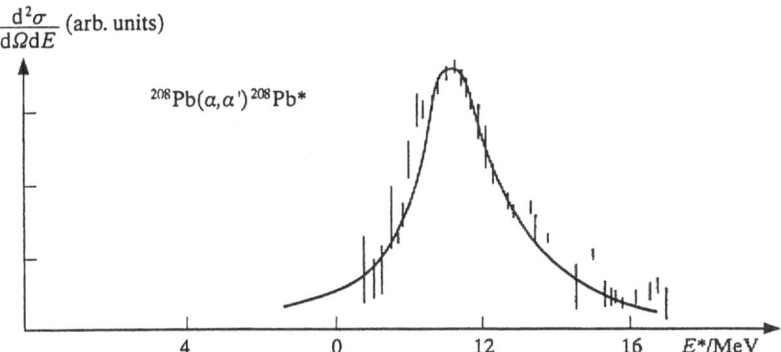

Fig. B.10. Multipole giant resonance in ^{208}Pb nucleus [B.4]. *Bars* – from data on (α, α') reaction; *solid curve* – theoretical calculation

The properties of nuclei are manifested in their spectroscopic character-istics such as the spin excitation energy and the parity of nuclear states, β- and γ-transition probabilities, the decay probabilities, including those for the production of α particles and of other light fragments (and, in recent years, of composite "clusters", also), decay branching ratios, fission param-eters, etc. The scheme of levels for each individual nucleus indicating the corresponding spectroscopic characteristics may serve as its "identity card". Such information updated in accordance with new data provided by exper-iments and theoretical publications are published regularly in the scientific literature. Consider the example of the scheme of levels of the ^{10}B nucleus, shown in Fig. B.11. The figures at the right $(3^+, 0; 1^+, 0; 0^+, 1;$ etc.) are the spins, parities, and isospins of the J^π, T state; the figures to the left (0.718, 1.740, etc.) are the excitation energies in MeV. The smaller the isospin of a nuclear state, the more favorable is a low value of its excitation energy – this is a general rule for all nuclei. In the nucleus considered, the "gap" between the lowest state with $T = 0$ (the ground state of the nucleus) and the lowest state with $T = 1$ is small, only about 1.5 MeV. This is due to the ^{10}B nucleus being one of the rare stable odd–even nuclei: the properties of the lowest levels of such nuclei are mainly determined by the pair interaction between the odd proton and the odd neutron (like in the deuteron, the np interac-tion in a composite nucleus is stronger for $T = 0$ than for $T = 1$), whereas the properties of these levels weakly reflect the interaction dynamics between the nucleons present in the even–even "core" of the nucleus. The picture in even–even nuclei is different; thus, for instance, the excitation energies of the lowest $T = 1$ levels of the ^{12}C, ^{24}Mg, ^{40}Ca nuclei amount to 12.00, 9.515, and 7.66 MeV, respectively.

In Fig. B.11, the thresholds are indicated for disintegration of the nucleus via various channels: 4.460 MeV for the channel ^{10}B $\rightarrow^6$ Li $+ \alpha$, 6.026 MeV for the channel ^{10}B $\rightarrow^8$ Be $+$ d, 6.585 MeV for the channel ^{10}B $\rightarrow^9$ Be $+$ p and so on (note that the proton and neutron channels are not the most favorable

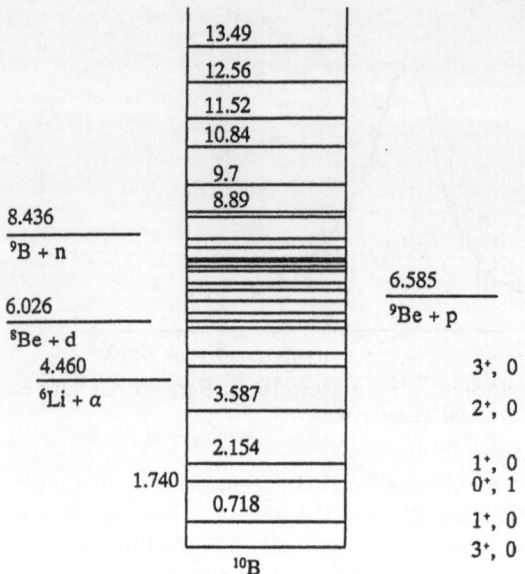

Fig. B.11. Scheme of levels in ^{10}B nucleus

from the point of view of energy). The cross sections are sometimes also written out, to the right and to the left of the scheme of levels, for those reactions in which this nucleus is produced as an intermediate compound system; the resonances in the cross sections yield information on the energy positions and quantum numbers of the levels of the nucleus dealt with. For example, the resonances in the reaction $^6\text{Li} + \alpha \rightarrow ^{10}\text{B} + \gamma$ correspond to levels of the ^{10}B nucleus with $T = 0$, and in the reaction $^7\text{Li} + ^3\text{He} \rightarrow ^{10}\text{B} + \gamma$ they correspond to levels with $T = 0$ and $T = 1$.

^{10}B is a light nucleus. In Fig. B.12, the scheme of levels is presented for the medium-heavy ^{56}Fe nucleus, which gives an idea of the character of the available spectroscopic information on such nuclei. In heavy nuclei the distances between levels are even smaller, the level density rises, and even the number of discrete levels that are below the nucleon escape threshold turns out to be enormous. Here, it is impossible to avoid a statistical approach with its specific concepts of nuclear temperature, of correlations between various properties of the levels, without the concept of evaporation that serves as the decay mechanism of the nucleus.

We have discussed certain properties of the nucleus from a traditional standpoint, considering it to be composed of protons and neutrons. Modern studies in the physics of the nucleus go much farther – toward a unified picture of nucleonic and subnucleonic degrees of freedom of nuclei. Much experimental material has been accumulated on the properties of "exotic" nuclei. Among these are hypernuclei, which, besides protons and neutrons, also contain Λ or Σ hyperons, and Δ nuclei, one of the nucleons of which is in the Δ (1232) excited state. Such "exotic" systems are of very great interest

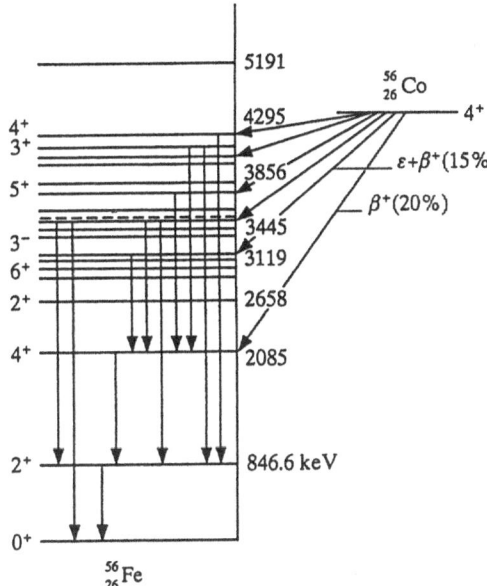

Fig. B.12. Scheme of levels in ^{56}Fe nucleus

both for investigating the baryon–baryon interaction (those of its aspects, for instance, which are out of reach if studies are restricted to ordinary nuclei), and for studies of the fundamental properties of nuclear matter. For example, a Λ hyperon in a hypernucleus exhibits a unique capability of penetrating the deepest layers of the nuclear core, right down to its central part, in which, according to existing theoretical ideas, the physical conditions should be close to the conditions in infinite nuclear matter.

C. Atoms and Molecules

The simplest atomic systems are the hydrogen atom and hydrogen-like ions. The spectrum of energy levels for electrons in the Coulomb field of a point-like charge,

$$V_{\mathrm{c}}(r) = -\frac{Ze^2}{r}, \tag{C.1}$$

is given by the Bohr formula

$$E_n = -\frac{\varepsilon_0 Z^2}{2n^2}, \qquad n = 1, 2, \ldots, \tag{C.2}$$

where

$$\varepsilon_0 = \frac{m_e e^4}{\hbar^2} = 27.21\,\mathrm{eV} \tag{C.3}$$

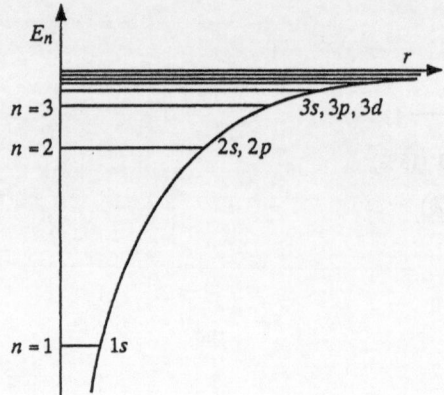

Fig. C.1. Diagram of levels in hydrogen atom and hydrogen-like ions

is the atomic unit of energy. The level spectrum (C.2) is depicted in Fig. C.1. To each level E_n there correspond n^2 states $|nlm\rangle$ ("quantum orbits" of the electron) with the wave functions

$$\varphi_{nlm}(r, \vartheta, \varphi) = R_{nl}(r) Y_{lm}(\vartheta, \varphi),$$
$$l = 0, 1, \ldots, n-1,$$
$$m = -l, \ldots, l.$$
(C.4)

We now present the explicit form of the normalized radial wave functions of the lowest states of the hydrogen atom and hydrogen-like ions:

$$R_{1s}(r) = \left(\frac{Z}{a_0}\right)^{3/2} 2 \exp\left(-\frac{Z}{a_0}r\right),$$

$$R_{2s}(r) = \left(\frac{Z}{2a_0}\right)^{3/2} \left(2 - \frac{Z}{a_0}r\right) \exp\left(-\frac{Z}{2a_0}r\right),$$
(C.5)

$$R_{2p}(r) = \left(\frac{Z}{2a_0}\right)^{3/2} \frac{Zr}{\sqrt{3}a_0} \exp\left(-\frac{Z}{2a_0}r\right),$$

where

$$a_0 = \frac{\hbar^2}{m_e e^2} = 0.53 \times 10^{-8}\,\text{cm}$$
(C.6)

is the atomic unit of length (the "Bohr radius"). The formulae for the mean values of a series of quantities characterizing the spatial dimensions and shapes of electron quantum orbits are also useful:

$$\langle r \rangle_{nl} = \frac{a_0}{2Z}\left[3n^2 - l(l+1)\right];$$

$$\langle r^2 \rangle_{nl} = \frac{a_0^2 n^2}{2Z^2}\left[5n^2 + 1 - 3l(l+1)\right];$$
(C.7)

$$\langle (r - \bar{r})^2 \rangle = \frac{1}{4}\left(\frac{a_0}{Z}\right)^2 \left[n^2(n^2 + 2) - l^2(l+1)^2\right].$$

The last formula, for example, demonstrates that the deviation of the radius of an orbit from its mean value is minimal for a given n if l is maximal, i.e., if $l = n - 1$; such orbits are termed "circular".

We now give the expressions relating the atomic units ε_0 and a_0 to other physical constants:

$$\varepsilon_0 = \frac{e^2}{a_0}, \tag{C.8}$$

$$\varepsilon_0 = m_e c^2 \left(\frac{e^2}{\hbar c}\right)^2 = \alpha^2 m_e c^2, \tag{C.9}$$

where $\alpha = e^2/\hbar c = 1/137$ is the fine structure constant;

$$a_0 = \frac{\hbar}{m_e c}\frac{\hbar c}{e^2} = \frac{\lambda_e}{\alpha}, \tag{C.10}$$

where $\lambda_e = \hbar/m_e c = 3.85 \times 10^{-11}\,\text{cm}$ is the Compton wavelength of the electron;

$$v_B = \frac{e^2}{\hbar} = \alpha c = 2.19 \times 10^8\,\text{cm/s} \tag{C.11}$$

is the Bohr velocity of the electron.

As we see from relations (C.2)–(C.7), the characteristics of electron states in hydrogen-like ions are totally similar to those of the electron in the hydrogen atom. Only the scales of the respective physical quantities change: the energy scale is "stretched out" in proportion to Z^2, the scale of linear dimensions is "compressed" in accordance with the $1/Z$ law. To be sure, this takes place at relatively small Z, when the Coulomb potential (C.1) can be considered to exhaust all the interaction between the electron and the nucleus. Indeed, the higher Z is, the more pronounced is the spin–orbit interaction

$$\hat{V}_{sl} = \frac{\hbar^2}{2m_e^2 c^2}\frac{1}{r}\frac{\partial V_c}{\partial r}(sl), \tag{C.12}$$

resulting in the "fine structure", i.e., in the energy levels of the electron splitting relative to the total angular momentum $j = l \pm 1/2$:

$$\Delta E_{nlj} = -\frac{\alpha^2 Z^4}{2n^3}\varepsilon_0 \left(\frac{1}{j + 1/2} - \frac{3}{4n}\right). \tag{C.13}$$

The dependence of the spin–orbit splitting on Z is different from (stronger than) that of the distance between levels of differing n; therefore, in the case of large Z the similarity of the level spectra in hydrogen-like ions and the spectrum of the hydrogen atom is noticeably violated. The fine structure of the $n = 2$ level of the hydrogen atom is shown in Fig. C.2. The distance

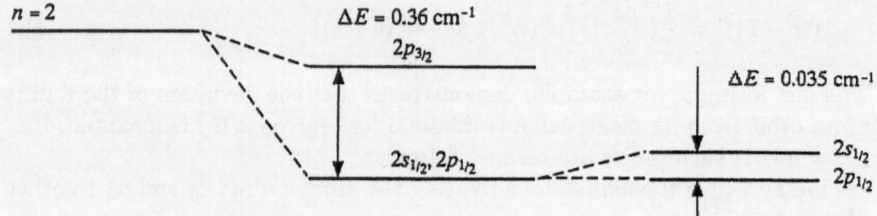

Fig. C.2. Fine structure of $n = 2$ level and Lamb shift of $2s_{1/2}$ level of hydrogen atom

between the components of the doublet is given in inverse centimeters. The following formulae are convenient for converting to usual units:

$$1\,\mathrm{cm}^{-1} \to 1.24 \times 10^{-4}\,\mathrm{eV},$$
$$1\,\mathrm{eV} \to 8.065 \times 10^{3}\,\mathrm{cm}^{-1}. \qquad (\text{C}.14)$$

The spin–orbit interaction potential of the electron (C.12) is "inherent" in the relativistic wave equation (the Dirac equation) for an electron moving in the symmetric central Coulomb field $V(r)$ of a nucleus, and from the point of view of nonrelativistic quantum theory (the Schrödinger equation) it represents a relativistic correction to the potential (C.1). Quantum electrodynamics permits us to take into account even more subtle effects. Its great achievement was the explanation of the Lamb shift (Fig. C.2). In hydrogen-like ions with large Z the level shift due to the finite size of the nucleus, i. e., due to deviation at small distances of the interaction potential between an electron and the nucleus from the Coulomb law (Fig. C.3), turns out to be significant. The deviations mentioned for the Bohr picture of the atom turn out to be much more significant in the so-called "exotic" atoms, where the role of the atomic electron is assumed by a negative muon, pion, kaon

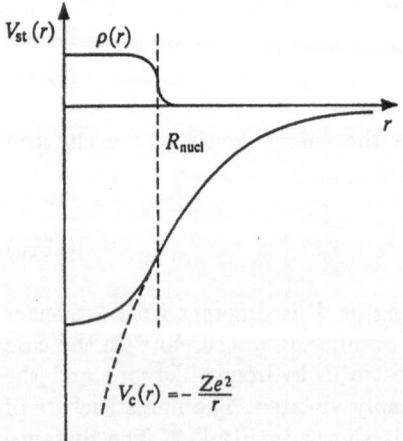

Fig. C.3. Electrostatic interaction potential, $V_{st}(r)$, of an electron interacting with a nucleus with account of the finite size of the nucleus [$\rho(r)$ is the charge distribution density in the nucleus]

Table C.1. D.I.Mendeleev Periodic table of elements. For each element the atomic number Z, the configuration of the valence shell, and the ionization potential (in eV) are given [C.1]

1	2	3	4	5	6	7	8	9	10	11	12	13	14	15	16	17	18
1H $1s$ 13.59																	2He $1s^2$ 24.59
3Li $1s$ 5.39	4Be $2s^2$ 9.32											5B $2s^2 2p$ 8.30	6C $2s^2 2p^2$ 11.26	7N $2s^2 2p^3$ 14.53	8O $2s^2 2p^4$ 13.62	9F $2s^2 2p^5$ 17.42	10Ne $2s^2 2p^6$ 21.56
11Na $3s$ 5.14	12Mg $3s^2$ 7.65											13Al $3s^2 3p$ 5.99	14Si $3s^2 3p^2$ 8.15	15P $3s^2 3p^3$ 10.49	16S $3s^2 3p^4$ 10.36	17Cl $3s^2 3p^5$ 12.97	18Ar $3s^2 3p^6$ 15.76
19K $4s$ 4.34	20Ca $4s^2$ 6.11	21Sc $3d 4s^2$ 6.56	22Ti $3d^2 4s^2$ 6.82	23V $3d^3 4s^2$ 6.74	24Cr $3d^5 4s$ 6.77	25Mn $3d^5 4s^2$ 7.43	26Fe $3d^6 4s^2$ 7.90	27Co $3d^7 4s^2$ 7.86	28Ni $3d^8 4s^2$ 7.63	29Cu $3d^{10} 4s$ 7.72	30Zn $3d^{10} 4s^2$ 9.39	31Ga $4s^2 4p$ 6.00	32Ge $4s^2 4p^2$ 7.90	33As $4s^2 4p^3$ 9.79	34Se $4s^2 4p^4$ 9.75	35Br $4s^2 4p^5$ 11.81	36Kr $4s^2 4p^6$ 14.00
37Rb $5s$ 4.19	38Sr $5s^2$ 5.69	39Y $4d 5s^2$ 6.22	40Zr $4d^2 5s^2$ 6.84	41Nb $4d^4 4s$ 6.88	42Mo $4d^5 6s$ 7.10	43Tc $4d^5 5s^2$ 7.28	44Ru $4d^7 5s$ 7.37	45Rh $4d^8 6s$ 7.47	46Pd $4d^{10}$ 8.34	47Ag $4d^{10} 5s$ 7.57	48Cd $4d^{10} 5s^2$ 8.99	49In $5s^2 5p$ 5.79	50Sn $5s^2 5p^2$ 7.34	51Sb $5s^2 5p^3$ 8.61	52Te $5s^2 5p^4$ 9.01	53I $5s^2 5p^5$ 10.45	54Xe $5s^2 5p^6$ 12.13
55Cs $6s$ 3.89	56Ba $6s^2$ 5.21	57-71	72Hf $5d^2 6s^2$ 6.80	73Ta $5d^3 6s^2$ 7.89	74W $5d^4 6s^2$ 7.98	75Re $5d^5 6s^2$ 7.88	76Os $5d^6 6s^2$ 8.73	77Ir $5d^7 6s^2$ 9.05	78Pt $5d^9 6s$ 8.96	79Au $5d^{10} 6s$ 9.23	80Hg $5d^{10} 6s^2$ 10.44	81Tl $6s^2 6p$ 6.11	82Pb $6s^2 6p^2$ 7.42	83Bi $6s^2 6p^3$ 7.29	84Po $6s^2 6p^4$ 8.42	85At $6s^2 6p^5$ 9.0	86Rn $6s^2 6p^6$ 10.75
87Fr $7s$ 4.0	88Ra $7s^2$ 5.28	89-103															

Lanthanide Series

57La $5d 6s^2$ 5.58	58Ce $4f 5d 6s^2$ 5.54	59Pr $4f^3 6s^2$ 5.47	60Nd $4f^4 6s^2$ 5.52	61Pm $4f^5 6s^2$ 5.58	62Sm $4f^6 6s^2$ 5.64	63Eu $4f^7 6s^2$ 5.67	64Gd $4f^7 5d 6s^2$ 6.15	65Tb $4f^9 6s^2$ 5.86	66Dy $4f^{10} 6s^2$ 5.94	67Ho $4f^{11} 6s^2$ 6.02	68Er $4f^{12} 6s^2$ 6.11	69Tm $4f^{13} 6s^2$ 6.18	70Yb $4f^{14} 6s^2$ 6.25	71Lu $4f^{14} 5d 6s^2$ 5.43

Actinide Series

89Ac $6d 7s^2$ 5.2	90Th $6d^2 7s^2$ 6.1	91Pa $5f^2 6d 7s^2$ 6.0	92U $5f^3 6d 7s^2$ 6.19	93Np $5f^4 6d 7s^2$ 6.27	94Pu $5f^6 7s^2$ 6.06	95Am $5f^7 7s^2$ 6.0	96Cm $5f^7 6d 7s^2$ 6.02	97Bk $5f^9 7s^2$ 6.23	98Cf $5f^{10} 7s^2$ 6.30	99Es $5f^{11} 7s^2$ 6.42	100Fm $5f^{12} 7s^2$ 6.5	101Md $5f^{13} 7s^2$ 6.6	102No $5f^{14} 7s^2$ 6.6	103Lr $5f^{14} 6d 7s^2$

Table C.2. Symbols for electron shells of atoms

Principal quantum number, n	Orbital and total angular momentum of electron, l_j						
	$s_{1/2}$	$p_{1/2}$	$p_{3/2}$	$d_{3/2}$	$d_{5/2}$	$f_{5/2}$	$f_{7/2}$
1	K						
2	L_{I}	L_{II}	L_{III}				
3	M_{I}	M_{II}	M_{III}	M_{IV}	M_{V}		
4	N_{I}	N_{II}	N_{III}	N_{IV}	N_{V}	N_{VI}	N_{VII}
5	O_{I}	O_{II}	O_{III}	O_{IV}	O_{V}	O_{VI}	O_{VII}

(μ^-, π^-, K^-) or an antiproton ($\bar{\mathrm{p}}$), or generally by a negatively charged heavy particle. This is due to the dimensions of an exotic atom being reduced (inversely proportional to the mass m_x of such a particle), as compared to the size of a usual atom.

We shall now turn to the description of atoms with many electrons. Their shell structure is reflected in the Mendeleev periodic table of the elements. Here, in Table C.1, we present it in an extended form, indicating the configurations of the valence shells and the ionization potentials of the atoms. Below, we shall take advantage of various sets of symbols used for indicating the electronic subshells of an atom; Table C.2 explains the meaning of one such set. Figure C.4 shows the binding energy of electrons on the inner shells of neutral atoms versus the atomic number Z. Comparison of the two upper (*solid* and *dotted*) curves reveals the screening of the Coulomb field of the nucleus to be significant even for the deepest K shell. Naturally, the screen-

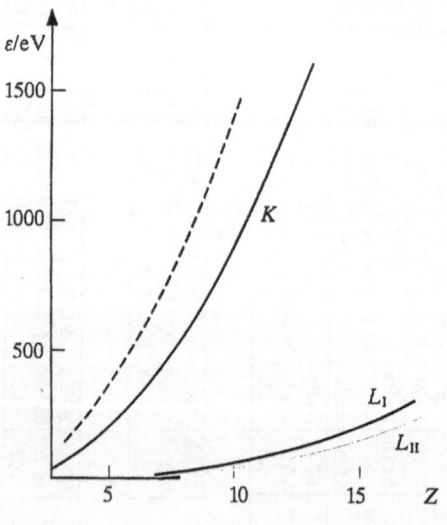

Fig. C.4. Binding energy of electrons of the inner shells K, L_{I}, L_{II} versus the atomic number Z. The *dotted line* shows the binding energy of the nonscreened K electron (C.2)

Table C.3. Effective charge of valence ns electron in atoms

Atom	$_3$Li	$_{11}$Na	$_{10}$K	$_{37}$Rb	$_{55}$Cs
Z_{eff}	1.26	1.84	2.26	2.77	3.21

ing effect is still stronger for the external electrons. In Table C.3, we present values of the effective charge Z_{eff} felt by the valence ns electrons in the ion cores of some alkali metal atoms. As one can see, Z_{eff} is considerably smaller than the charge of the nucleus, Z, although it differs quite significantly from unity, which would be expected in the case of extremely strong screening. Such extremely strong screening is only observed in the case of very highly excited so-called *Rydberg levels* of the valence electron (the range of the principal quantum numbers of Rydberg states amounts, at present, to $n = 100$ and higher).

Methods for the theoretical description of the structure of atoms and ions with many electrons are quite diverse. The variational method, various modifications of the Hartree–Fock method, and the Thomas–Fermi method are widely applied.

The Hartree–Fock method is the most universal method. It is consistent in taking the shell structure of an atom into account, good in representing the individual features of various atoms, and successful not only in determining the parameters of an atom in the ground state, but also in calculating the spectra of levels and the transition probabilities between them. Implementation of this method requires numerical solution of sets of coupled integro-differential equations, and it was this circumstance that for a long time imposed restrictions on its practical utilization. At present, the situation has undergone essential changes in this respect: the Hartree–Fock method is involved in many libraries of standard computer programs and is widely applied in various computations carried out both by theorists and experimentalists. Figure C.5 presents the example of the electron density distribution $\rho_{\text{el}}(r)$ in the argon atom calculated by the Hartree–Fock method; the density oscillations due to the contributions of individual electron shells are clearly seen.

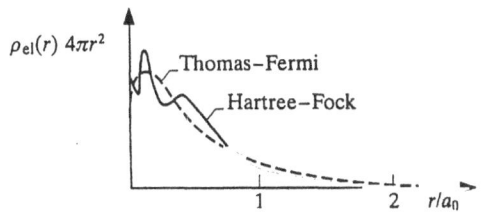

Fig. C.5. Radial electron density distribution $\rho_{\text{el}}(r)\,4\pi r^2$ in the argon atom

The electron density in the argon atom calculated by the Thomas–Fermi method is presented in the same picture. This method is more rough. The electron density distribution given by it only approximately traces the calculations by the Hartree–Fock method: no shell effects are seen, the behavior of the density in the peripheral region of the atom is described erroneously. However, an essential advantage of the Thomas–Fermi method is its simplicity. This method, based on a statistical approach, permits a unique formula to be obtained for the electron density distributions in various atoms:

$$\rho_{\text{el}}(r) = Z^2 f\left(\frac{r}{0.885 a_0\, Z^{-1/3}}\right). \tag{C.15}$$

Here a_0 is the Bohr radius and $f(x)$ is a certain universal function, which has been determined and tabulated by numerical solution of the Thomas–Fermi equation. From (C.15) it can be seen that as Z increases the electron density distributions in various atoms are similar, according to the Thomas–Fermi method, whereas the linear scale of the distribution undergoes "compression" following the $Z^{-1/3}$ law. For further use we also present here the expression for the mean electron velocity in the Thomas–Fermi atom:

$$\langle v_e \rangle \approx 0.7 \cdot Z^{2/3} v_{\text{B}}, \tag{C.16}$$

where v_{B} is the Bohr velocity.

We shall start the discussion of the spectra of atomic levels with the simplest case of alkali metal atoms. The lowest excited states of such atoms correspond to transition of the valence electron from the ground state ns to the state $n'l$ with $l \geq 0$; the ion core (all its shells are occupied) feels this

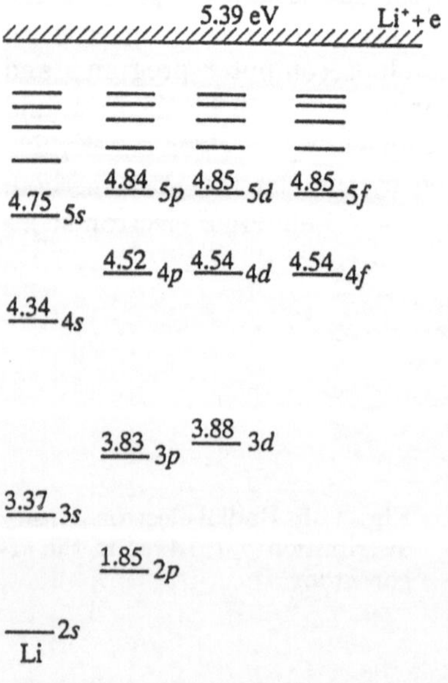

5.39 eV   Li$^+$+e

4.75 5s   4.84 5p   4.85 5d   4.85 5f

4.52 4p   4.54 4d   4.54 4f

4.34 4s

3.83 3p   3.88 3d

3.37 3s

1.85 2p

2s
Li

Fig. C.6. Level spectrum of the valence electrons in the Li atom

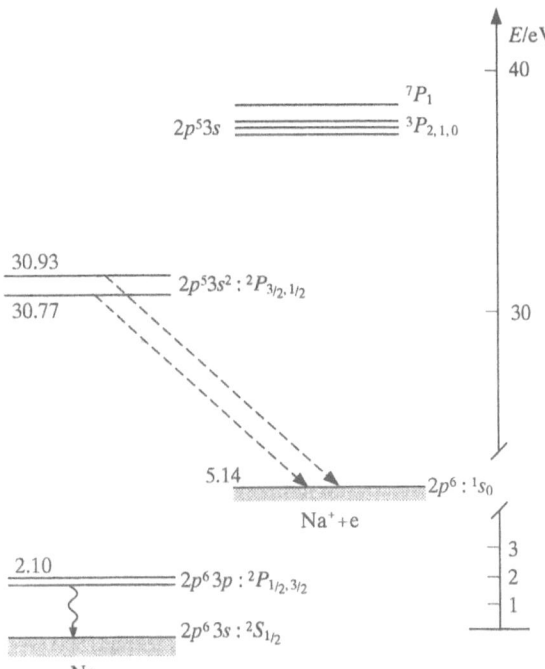

Fig. C.7. Position and decay schemes of the lower autoionization states of the Na atom

transition very weakly. The interaction potential of the valence electron with the ion core differs from the Coulomb potential, so no level degeneracy relative to l, which is peculiar to hydrogen, occurs (Fig. C.6). The levels concentrate toward the ionization threshold of the valence electron; above this threshold there are levels that correspond to the excitation of electrons of the core (Fig. C.7). These are the so-called autoionization levels, or *autoionization states of the atom*: sufficient excitation energy is accumulated in these atomic states for spontaneous ionization to take place:

$$A^* \rightarrow A^- + e. \tag{C.17}$$

The inherent widths Γ of atomic autoionization levels may amount to some tenths of an eV; this corresponds to an autoionization state having a mean lifetime of the order of 10^{-14} s. When autoionization occurs, the states of at least two electrons change: one undergoes transition from a bound state to the continuous spectrum, and the other undergoes transition from a higher to a lower level, thus providing the first electron with the necessary energy. Thus, the process of autoionization is akin to the *Auger process* taking place when vacancies in the inner shells of the atom are being filled up. Both processes are due to the existence of pair correlations between the electrons in atoms.

The effect of pair correlations is clearly seen in the spectra of atomic levels of alkaline earth elements: Be, Mg, Ca, and so on. The main configu-

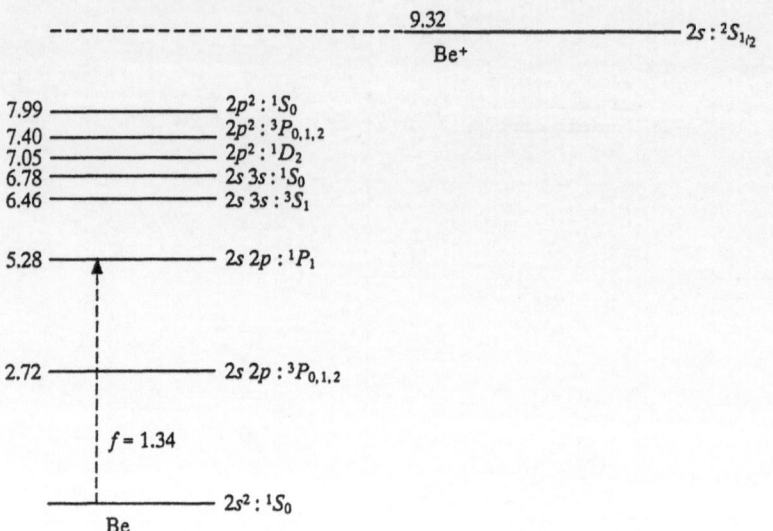

Fig. C.8. Diagram showing the lowest levels of the beryllium atom

ration of such atoms is $(ns)^2$, whereas the configuration of the lowest excited states is (ns, np) or $(np)^2$. Now, consider the positions of the $2s2p$: ^1P and $2s2p$: ^3P levels in the Be atom (Fig. C.8). In the singlet state ^1P the wave function of the two external electrons, $\Psi_{2s2p}(r_1, r_2)$ is symmetric with respect to permutation of their space coordinates, in the triplet states ^3P it is antisymmetric. Therefore, the mutual repulsion of these electrons acts stronger in the first case than in the second, and the $2s2p$: $^1P_1$ level turns out to have a significantly higher atomic excitation energy than do levels of the same configuration $2s2p$: $^3P_{0,1,2}$.

Now, consider the issue of electromagnetic transitions in atoms. Unlike in the nuclear case, the photon wavelength here always greatly exceeds the dimensions of the system, and, consequently, electric dipole transitions are sure to dominate over all other transitions. Such an $E1$ transition (in atomic physics they are termed *optically allowed*) is shown in Fig. C.8 by the dotted arrow. The symbol f by this arrow is the strength of the transition oscillators. Generally, the oscillator strength for the transition $|a\rangle \rightarrow |b\rangle$ is expressed in terms of the matrix element of the dipole electric moment operator of the atom, $D = \sum_{j=1}^{Z} ez_j$, and the transition energy $\omega_{ba} = \frac{1}{\hbar}(\varepsilon_b - \varepsilon_a)$:

$$f_{ba} = \frac{2m_e \omega_{ba}}{\hbar e^2} \left| \langle b|\hat{D}|a\rangle \right|^2. \qquad (C.18)$$

The same two quantities, $|\langle b|\hat{D}|a\rangle|$ and ω_{ba}, but within another combination yield the probability of spontaneous electromagnetic transition occuring between the levels considered (the partial component of the radiation width $\Gamma_{b\rightarrow a}$ of the upper one of these levels):

$$\omega_{b \to a} = \frac{\Gamma_{b \to a}}{\hbar} = \frac{4}{3} \frac{\omega_{ba}^3}{\hbar e^3} \left| \left\langle b \left| \hat{D} \right| a \right\rangle \right|^2. \tag{C.19}$$

The advantage of the combination (C.18), as compared with (C.19), consists in that the sum of the oscillator strengths for transitions from a given state $|a\rangle$ to all possible states $|b\rangle$ satisfies a strict but, at the same time, very simple relation – the *sum rule*:

$$\sum_b f_{ba} = Z, \tag{C.20}$$

where Z is the number of electrons in the atom. Thus, the oscillator strength for the transition $|a\rangle \to |b\rangle$ shows the relative intensity of this transition among all possible transitions from the intial state $|a\rangle$ under consideration. The transition $(2s)^2 : S_0 \to 2s2p : {}^1P_1$ in the beryllium atom indicated in Fig. C.8 pertains to allowed strong optical transitions; here the oscillator strength equals 1.34, while the total sum $\sum_b f_{ba}(\mathrm{Be}) = 4$.

The distribution of oscillator strengths over the excitation spectrum of the atom differs in different atoms. In the alkali metal atoms the strong transitions $ns_{1/2} \to np_{1/2}; ns_{1/2} \to np_{3/2}$ are to be singled out; they can be roughly interpreted as one-particle transitions having no influence on the electrons in the occupied shells of the ion core. Indeed, the respective oscillator strengths for such transitions amount to the following: 0.247 and 0.494 in Li; 0.318 and 0.637 in Na; 0.35 and 0.70 in K; 0.32 and 0.67 in Rb; 0.39 and 0.81 in Cs; the sum of each pair is close to unity. We could also have included hydrogen in this series but it must be noted that the oscillator strength of the corresponding transition is significantly lower than unity here: $f(1s \to 2p) = 0.416$. Transitions to the states $3p$, $4p$, and so on are weaker and only complement the $1s \to 2p$ insignificantly, so half of all the strength oscillators fall within the entire discrete spectrum, whereas the other half involve transitions to the continuous spectrum, i. e., ionization.

To conclude this introductory section we shall consider certain issues of molecular structure. The simplest molecule is the hydrogen molecular ion $\mathrm{H_2^+}$, which represents a three-particle system composed of two protons and an electron. It will be convenient to take advantage of its example for examining a number of essential points concerning the nature of chemical bonds and the character of molecular excitation.

Classical physics is not capable of explaining why the interaction of a hydrogen atom with a second proton may result in the production of a bound state of the three particles forming the $\mathrm{H_2^+}$ ion. This problem is resolved in quantum theory. We shall adopt the adiabatic approximation: since the nuclei present in the molecule, being heavy particles, move much slower than the electron, it is possible to introduce the concept of energy levels of the electron, $\varepsilon_n(R)$, for fixed positions of the nuclei and to determine these levels, having first defined the Hamiltonian

$$\hat{H}_{\mathrm{el}} = -\frac{\hbar^2}{2m_{\mathrm{e}}} \nabla^2 - \frac{e^2}{r_{\mathrm{A}}} - \frac{e^2}{r_{\mathrm{B}}} \tag{C.21}$$

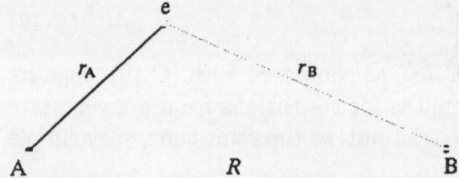

Fig. C.9. Coordinates of particles in hydrogen molecular ion

(the meaning of the notation is seen from Fig. C.9). The electron wave function in the ground state of the molecular ion H_2^+ is approximated by a symmetric superposition of two nonperturbed wave functions of the isolated atom $\psi_{1s}(r) = 1/\sqrt{\pi a^3}e^{-r/a}$:

$$\Psi_e(R, r) \sim \Psi_{1s}(r_A) + \Psi_{1s}(r_B),\qquad (C.22)$$

and the dependence of the electron energy on the distance between the nuclei is given by the expression

$$\varepsilon(R) = \varepsilon_{1s} + \frac{\left\langle \Psi_{1s}(r_A)\left| -\frac{e^2}{r_B}\right|\Psi_{1s}(r_A)\right\rangle + \left\langle \Psi_{1s}(r_A)\left| -\frac{e^2}{r_B}\right|\Psi_{1s}(r_B)\right\rangle}{1 + \left\langle \Psi_{1s}(r_A)|\Psi_{1s}(r_B)\right\rangle}.\qquad (C.23)$$

The first term in the numerator has a simple physical meaning: it is the mean interaction energy of an electron in the $1s$ state in the field of proton A in the presence of the "alien" proton B at a distance R from A. The second term in this numerator cannot be understood within the framework of classical physics: it reflects the quantum effect of interference between the probability amplitudes for the electron to be found in the vicinity of proton A and in the vicinity of proton B. Without taking into account this term, no stable H_2^+ system can be obtained.

The energy of the electron $\varepsilon(R)$ together with the energy of Coulomb repulsion between the nuclei, e^2/R, determines, in the adiabatic approximation, the potential energy of the interaction between the nuclei:

$$U_{\text{eff}}(R) = \frac{e^2}{R} + \varepsilon(R).\qquad (C.24)$$

Its dependence on the distance between the nuclei is shown in Fig. C.10. The minimum of potential energy (C.24) is found at $R \approx 2.5a = 1.32 \times 10^{-8}$ cm, which somewhat exceeds the equilibrium distance of 1.06×10^{-8} cm between the nuclei in the H_2^+ molecular ion.

We shall follow the motion of the nuclei with the aid of the stationary Schrödinger equation for their wave function $\Psi(R, \vartheta, \varphi)$, where the angles ϑ and φ indicate the orientation of the internuclear axis of the molecule; here we shall use the reduced mass of the system

$$\mu = \frac{m_A m_B}{m_A + m_B}.\qquad (C.25)$$

Since the effective potential energy depends on only R, the problem reduces to the radial Schrödinger equation

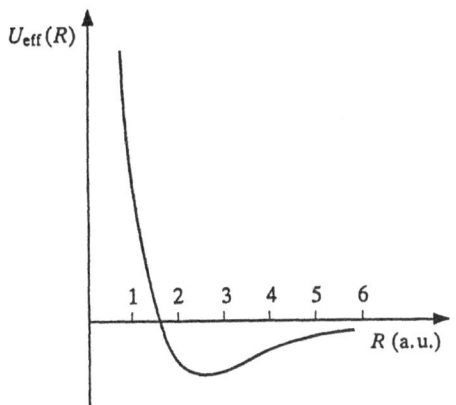

Fig. C.10. Effective potential interaction energy between protons in hydrogen molecular ion

$$\frac{d^2}{dR^2}[RR_J(R)] + \frac{2\mu}{\hbar^2}\left[E - U_{\text{eff}}(R) - \frac{\hbar^2 J(J+1)}{2\mu R^2}\right]RR_J(R) = 0, \quad (C.26)$$

from which we find the set of eigenvalues E_{vJ} for each J value of the angular momentum of the molecule. Here, the wave function $\Psi_{vJM}(R, \vartheta, \varphi)$, describing the motion of the nuclei, is the product of factors, one of which is related to rotation of the molecule, whereas the other, R_J, is related to oscillations of the nuclei along its axis:

$$\Psi_{vJM}(R, \vartheta, \varphi) = R_{vJ}(R)Y_{JM}(\vartheta, \varphi). \quad (C.27)$$

In accordance with the adiabatic approximation, this wave function itself is a factor in the complete wave function of the molecule:

$$\Psi(r, R, \vartheta, \varphi) = \Psi_e(r, R)R_{vJ}(R)Y_{JM}(\vartheta, \varphi). \quad (C.28)$$

The presence of $\hbar^2 J(J+1)/2\mu R^2$ in (C.26) results in the equilibrium distance between the nuclei depending on their rotational motion, while the mean linear dimensions of the molecule increase with the angular momentum J. Now, let us expand the sum of the potential energy and the centrifugal energy operators in a series in powers of the deviation from the equilibrium value R_J:

$$U_{\text{eff}}(R) + \frac{\hbar^2 J(J+1)}{2\mu R^2} = \left[U_{\text{eff}}(R) + \frac{\hbar^2 J(J+1)}{2\mu R^2}\right]_{R=R_J}$$

$$+ \frac{1}{2}\left[U_{\text{eff}}(R) + \frac{\hbar^2 J(J+1)}{2\mu R^2}\right]''_{R=R_J}(R - R_J)^2 + \dots. \quad (C.29)$$

If the higher-order terms in this expansion are dropped, then we arrive at an equidistant spectrum, characteristic of the harmonic oscillator, for vibrational molecular excitations:

$$E_{Jv} = \text{const} + \frac{\hbar^2 J(J+1)}{2\mu R_0^2} + \hbar\omega_0(v+1/2) + \Delta E_{Jv}, \quad (C.30)$$

where ω_0 is the oscillation frequency, and $v = 0, 1, 2, \ldots$ is the vibrational quantum number. There exists an interaction between the rotation and vibration of nuclei in a molecule. In (C.30), it is taken into account by the last term

$$\Delta E_{Jv} \sim \frac{\hbar^2 J(J+1)}{\mu^2 R_0^4 \hbar \omega}(v + 1/2). \tag{C.31}$$

The problem of the hydrogen molecular ion is of great practical importance, for example, in dealing with a number of mesomolecular phenomena (see Lecture 14). On the other hand, the example of the simplest molecular complex allows the tracing of the general principles of the theoretical description of molecules, including, also, molecules consisting of many atoms. Although in the particular case of the two-atom molecule the energy of the electrons $\varepsilon_n(R)$ depends, in the adiabatic approximation, on the sole parameter R, in the case of a molecule composed of N nuclei $(N > 2)$ it depends on $3N - 6$ independent variables characterizing the mutual positions of nuclei in the molecule.

The symmetry of a molecule determines the symmetry of the mean field in which its electrons move and, consequently, the character of the spectrum of its electron excitations. At the same time, the symmetry of the molecule also determines the possible types of its oscillations (effective methods for classifying vibrational excitations of molecules on the basis of their symmetry properties are provided by group theory). The mutual arrangement of nuclei in a molecule also determines its moments of inertia relative to various axes, which are the main parameters characterizing rotational excitation of the molecule.

The example of the hydrogen molecular ion was used above to demonstrate the existence of a relationship between the oscillatory and rotational excitations of molecules. Such a relationship also exists in the general case: small rotational excitations are imposed on the more energetic oscillations of a molecule, which leads to the line-band picture in the infrared spectra of molecules. There also exists a characteristic relationship between electron excitations and molecular oscillations. This relationship is manifested, for instance, in the *Frank–Condon principle*. This has to do with electron transitions in molecules. In Fig. C.11 two electron states of a two-atom molecule are depicted schematically: $\varepsilon_0(R)$ and $\varepsilon_1(R)$. The average distances between the nuclei in states $|1\rangle$ and $|2\rangle$ are shifted relative to each other. Assume the molecule to be in the lowest oscillatory state of the main term $\varepsilon_0(R)$. Under the influence of some external perturbation the electrons of the molecule change state and it undergoes transition to state $\varepsilon_1(R)$. The nuclei of the molecule have no time, then, to change their initial positions and, consequently, excitation of the electron state of the molecule is accompanied by excitation of its vibrations. From the schematic picture presented above it can be seen that if the spread $(\Delta R)_0$, connected to the vibrations of nuclei in the ground state, is sufficiently large, then, in the case of electron excitation

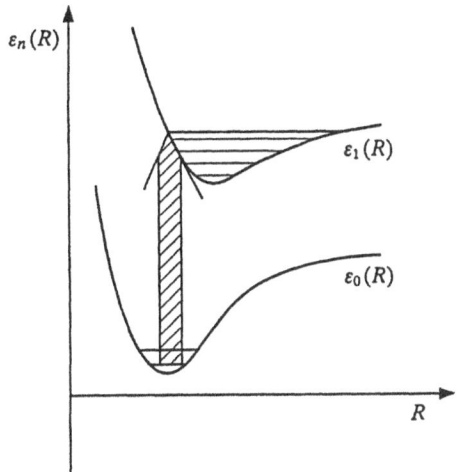

Fig. C.11. Scheme for origin of band of electron–vibrational transitions in molecules

having a noticeable probability, not only one, but several vibrational levels may be excited; there arises a band of electron–vibrational transitions in the molecule.

Table C.4 shows the main parameters characterizing the two-atom molecules H_2, N_2, and NaCl: the ionization potential I, the dissociation energy D, the equilibrium distance R_0, and the vibration frequency (energy) for the main term ε_0. It would be appropriate to add that the dissociation energy of the heavy hydrogen molecule D_2 differs, as shown by experiments, from the corresponding value for the H_2 molecule:

$$D(H_2) = 4.48\,\text{eV}; \qquad D(D_2) = 4.56\,\text{eV}.$$

This is a manifestation of the effect of vibration of the molecules in their ground state. The dissociation energy D of the molecule is the depth of the potential well minus the vibrational energy in the ground state:

$$D = \left|U_{\text{eff}}(R_0)\right| - \tfrac{1}{2}\hbar\omega_0. \tag{C.32}$$

The vibration frequency ω_0 is expressed through the rigidity coefficient k determining the behavior of the term (C.29) in the vicinity of the equilibrium point:

Table C.4. Main parameters of some diatomic molecules

Molecule	I/eV	D/eV	$R_0/\text{Å}$	$\omega_0/\text{cm}^{-1}\big/\hbar\omega_0/\text{eV}$
H_2	15.42	4.48	0.74	4400/0.55
N_2	15.58	9.76	1.12	2200/0.28
NaCl	7.64	4.3	2.36	365/0.045

$$U_{\text{eff}}(R) \approx U_{\text{eff}}(R_0) + \tfrac{1}{2}k(R - R_0)^2 + \ldots, \tag{C.33}$$

and the reduced mass of the molecule, μ:

$$\omega_0 = \sqrt{\frac{k}{\mu}}. \tag{C.34}$$

Hence it is seen that the vibration frequency ω_0 and, together with it, the vibration energy, $\tfrac{1}{2}\hbar\omega_0$, of the D_2 molecule are $\sqrt{2}$ times smaller than in the case of the H_2 molecule. Thus,

$$D(D_2) - D(H_2) = -\tfrac{1}{2}\hbar\omega_0(D_2) - \tfrac{1}{2}\hbar\omega_0(H_2)$$

$$= (1/2)(1 - 1/\sqrt{2})\hbar\omega_0(H_2). \tag{C.35}$$

Substituting the value of $\hbar\omega_0(H_2) = 0.55\,\text{eV}$, we obtain $D(D_2) - D(H_2) = 0.078\,\text{eV}$, fully in accordance with the experimental data presented above.

Part I

Interaction of Charged Particles with Matter

First Lecture

1.1 Ionization Stopping and Multiple Scattering of Fast Heavy Particles in Disordered Media: Relationship Between Parameters Characterizing the Passage of Particles Through Matter and Characteristics of Elementary Processes

When a fast charged particle passes through a material (gas, liquid, solid, plasma) it experiences multiple elastic and inelastic collisions with the atoms (molecules) of this material. To start with, we shall set aside all the consequences such collisions have for the medium through which the particle passes, and will be interested in only the passage of the particle.

Let $\langle \Delta E \rangle$ be the mean energy lost by the particle as it travels along a path segment Δx. In the case of small Δx the quantity $\langle \Delta E \rangle$ is proportional to Δx. Therefore, the rate at which the particle loses energy is conveniently characterized by the ratio $\langle \Delta E \rangle / \Delta x$ depending, in the limit $\Delta x \to 0$, on the mass, charge, and velocity of the particle and on the properties of the material in the vicinity of the point where the particle is found. In this limit the ratio is conventionally termed the *stopping power of the material* and is denoted by $-\mathrm{d}E/\mathrm{d}x$; this notation takes into account that the change in energy of a particle passing through a material is always a negative quantity (so $-\mathrm{d}E/\mathrm{d}x$ is a positive quantity). The stopping power introduced has the dimension of energy/length, i.e. MeV/cm, for example. The path covered by the particle in a material can be measured not only in centimeters or microns, but also by the mass of the layer covered, for example, in units of g/cm^2:

$$X = \rho x \,, \tag{1.1}$$

where ρ is the density of the medium. Connected with this is another, equivalent, definition of the stopping power:

$$-\frac{\mathrm{d}E}{\mathrm{d}X} = \frac{1}{\rho} \left(-\frac{\mathrm{d}E}{\mathrm{d}x} \right) \,. \tag{1.2}$$

This quantity is called the *mass stopping power*; its dimensionality is, for example, MeV/(g/cm^2). A decisive contribution to the stopping power of a material is due to inelastic collisions of a particle with atoms in the medium when the energy of the particle is spent on excitation or ionization of the atoms. Further, we shall see that ionization processes are more essential here than are excitation processes, which are not accompanied by the atoms losing their electrons. In this connection, the stopping of charged particles passing through a material is usually called *ionization stopping*, and the energy losses occurring in the stopping process are termed ionization losses. Energy loss per unit path, i.e., the stopping power of a material $-dE/dx$, is termed *ionization stopping power*.

Our task consists in expressing the stopping power of a material through the characteristics of elementary collision processes of a particle with individual atoms of the material. We shall adopt the *hypothesis of independent successive pair collisions* of a particle with atoms of the medium. When our medium is a gas, such a hypothesis seems quite valid, but the situation, however, is much more complicated if the particle passes through a condensed medium. The limitations of this hypothesis are especially clear when one considers the passage of charged particles through crystals (see Lecture 5). So, we shall adopt the hypothesis of independent pair collisions as a first approximation, and at the same time we shall consider the medium through which the particle passes to be homogeneous and totally disordered.

To reveal the essence of the elementary interaction processes of a particle with the atom of a material it is necessary to know the structure of the atom and its characteristics, for instance, the spectrum of its energy levels. Figure 1.1 presents a schematic picture of such a spectrum. While the excitation energy of the atom is lower than its ionization potential I, the spectrum of atomic levels remains discrete. Then comes the continuous spectrum, within the range of which certain individual spontaneously decaying (autoionization) states are encountered. We shall denote states pertaining to the discrete spectrum of an atom by the symbol $|n\rangle$ and write the corresponding wave functions in the form $\varphi_n(\xi) = \varphi_n(r_1\sigma_1, r_2\sigma_2, \ldots, r_z\sigma_z)$, where ξ represents the space and spin variables of all the Z electrons of the atom. The state vectors $|n\rangle$ are orthonormalized:

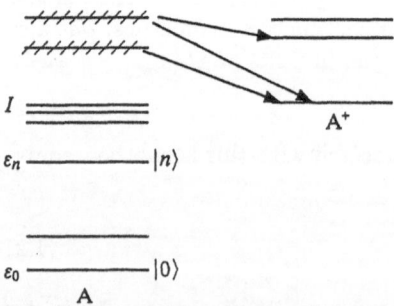

Fig. 1.1. Arrangement of discrete and autoionization levels of an atom (I is the ionization potential)

$$\langle n|n'\rangle = \delta_{nn'},\tag{1.3}$$

and together with the states of the continuum they constitute a complete set:

$$\sum_n \varphi_n(\xi)\varphi_n^*(\xi') + \ldots = \delta(\xi - \xi')\tag{1.4}$$

(here, the dots stand for the integral over the continuum). We shall also write the completeness condition (1.4) in the symbolic form

$$\sum_n |n\rangle\langle n| + \ldots = \hat{I},\tag{1.5}$$

where $\hat{I}$ is the unit operator in the Hilbert space of atomic states.

Now, consider the passage of a charged particle 'a' through a material, and let its mass significantly exceed the electron mass:

$$m_{\mathrm{a}} \gg m_{\mathrm{e}}\tag{1.6}$$

(the passage of electrons through a material is dealt with in Lecture 4). From a formal point of view our approach is valid for particles of abitrary charge Z_{a}, but in the case of multicharge ions ($Z_{\mathrm{a}} \gg 1$) one should actually take into account a number of special physical processes, which we shall not do for the present (see Sect. 4.5). Thus, we shall assume a represents one of the following: a proton p, an α particle, the nuclei of the lightest elements, and, muons ($\mu^{\pm}$), charged mesons ($\pi^{\pm}, \mathrm{K}^{\pm}$), and baryons (for instance, the Σ^-, Σ^+ hyperons and the antiproton $\bar{\mathrm{p}}$). Generally, when a particle undergoes collision with an atom A, various reaction channels may be realized:

$$
\begin{array}{lll}
\mathrm{a} + \mathrm{A} \longrightarrow \mathrm{a} + \mathrm{A} & - & \text{elastic scattering} \\[2pt]
\longrightarrow \mathrm{a}' + \mathrm{A}^* & - & \text{inelastic scattering} \\
 & & \text{with excitation of atom} \\[2pt]
\longrightarrow \mathrm{a}' + \mathrm{A}^* + \mathrm{e} & - & \text{inelastic scattering with} \\
 & & \text{ionization of atom} \\[2pt]
\longrightarrow\quad\quad & - & \text{other channels (charge} \\
 & & \text{exchange, bremsstrahlung, etc.)}
\end{array}\tag{1.7}
$$

We shall characterize the probability of a collision a + A occuring by the total effective cross section σ; it is the sum of the total elastic scattering cross section and the total cross section of inelastic processes:

$$\sigma = \sigma_{\mathrm{el}} + \sigma_{\mathrm{inel}}.\tag{1.8}$$

We shall call the effective cross sections σ_n of processes a + A $\longrightarrow$ a' + A$_n^*$, corresponding to the excitation of some particular levels of the atom, *partial inelastic scattering cross sections*. If charge exchange, bremsstralung, and other less important channels are not taken into account, the total cross section for inelastic processes may be written as

$$\sigma_{\text{inel}} = \sum_{n \neq 0} \sigma_n + \dots, \tag{1.9}$$

where the sum over n embraces all the excited discrete levels of the atom, and the dots imply integration over the continuum. We shall further write the relationsip (1.9) [and all relations similar to it, for example, (1.4) or (1.5)] simply in the form

$$\sigma_{\text{inel}} = \sum_{n \neq 0} \sigma_n, \tag{1.10}$$

implying summation over n to include, also, integration over the continuum. When it is necessary to distinguish between excitation processes of discrete levels and ionization of the atom, we shall explicitly decompose the inelastic scattering cross section into the respective terms:

$$\sigma_{\text{inel}} = \sigma_{\text{excit}} + \sigma_{\text{ioniz}}. \tag{1.11}$$

Having introduced the characteristics of the elementary collision process a + A, we shall go back to our task of expressing the stopping power of a material in terms of these characteristics.

Now, consider the passage of particle 'a' through a thin material layer consisting of atoms of sort A (Fig. 1.2). According to the adopted hypothesis of independent successive collisions, the ionization losses of the particle equal the sum of losses due to the excitation (ionization) of individual atoms. If a certain collision causes an atom to undergo a transition $|0\rangle \rightarrow |n\rangle$, then each of such collisions results in the particle losing an amount of energy equal to $\varepsilon_n - \varepsilon_0$ (strictly speaking, a small amount of the particle's energy is spent on the recoil of the atom as a whole, but this energy can usually be neglected). Consequently, if ΔN_n is the mean number of collisions undergone by the particle with atoms of the material in a layer Δx and resulting in a transition $|0\rangle \rightarrow |n\rangle$, then the mean change in the energy of the particle upon passage through the layer Δx is

$$\Delta E = - \sum_n (\varepsilon_n - \varepsilon_0) \Delta N_n. \tag{1.12}$$

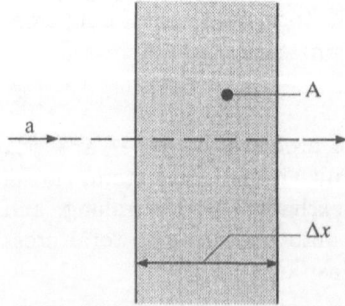

Fig. 1.2. Passage of a fast charged particle through a layer Δx of a substance

It now remains for us to express ΔN_n in terms of the partial cross sections σ_n (independent of energy):

$$\Delta N_n = n_0 \sigma_n \Delta x, \tag{1.13}$$

here n_0 is the number of atoms of sort A in a unit volume of the material.[1] Substituting (1.13) into (1.12) and going to the limit $\Delta x \to 0$, we obtain

$$-\frac{\mathrm{d}E}{\mathrm{d}x} = n_0 \sum_n (\varepsilon_n - \varepsilon_0) \sigma_n. \tag{1.14}$$

Thus, the stopping power $-\mathrm{d}E/\mathrm{d}x^2$ is the product of two factors. One of them, the number n_0 of atoms of sort A per unit volume, is in no way related to the properties and parameters of the particle we are interested in. On the contrary, the other factor is independent of the density of the material but contains all the information concerning the interaction of particles 'a' and 'A'. Its standard notation is

$$S(E) \equiv \sum_n (\varepsilon_n - \varepsilon_0) \sigma_n. \tag{1.15}$$

This is the so-called *effective stopping*. According to (1.14) the effective stopping $S(E)$ is the stopping power of a material reduced to the unit density of the number of atoms in the material:

$$-\frac{\mathrm{d}E}{\mathrm{d}x} = n_0 S(E). \tag{1.16}$$

Stopping represents only one aspect of the passage of a particle through matter. At the same time, each collision event is accompanied by some change in the direction of motion of the particle. Let us consider how a particle beam initially having a certain direction "spreads out" due to multiple collisions of each of the beam particles with atoms of the material.

We shall introduce two characteristics and, consequently, two different notations: Θ_i, the deviation angle of the particle from the initial direction after the ith collision; ϑ_i, the scattering angle resulting from the ith collision. Let v_i be the velocity vector of the particle after the ith collision (accordingly, v_{i-1} is the velocity vector before the ith collision, i.e., after the $(i-1)$th collision). All these quantities are shown in Fig. 1.3.

Let vector v_{i-1} lie in the plane of the picture, then vector v_i will not, generally speaking, be in this plane, and, consequently, its position will be characterized by the azimuthal angle φ_i as well as by the scattering angle ϑ_i.

From geometry it follows that

$$\cos \Theta_i = \cos \Theta_{i-1} \cos \vartheta_i + \sin \Theta_{i-1} \sin \vartheta_i \cos \varphi_i. \tag{1.17}$$

---

[1] The effective collision cross section σ of two particles, $a + A \to f$ is defined as the mean number of events per second, f, reduced to a *single* target particle A and to a *unit* flux j_a of incident particles. It is readily seen that relation (1.13) is consistent with this definition.

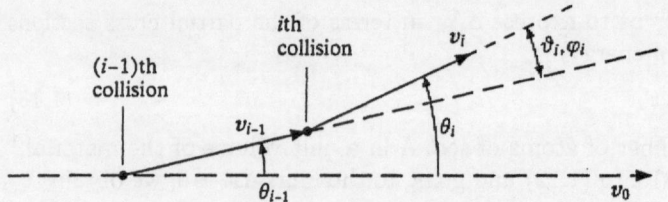

Fig. 1.3. Direction of motion of a particle after the $(i-1)$th and ith collisions

Thus, we have expressed the deviation angle of the particle from the initial direction after the ith collision via the deviation angle Θ_{i-1} after the preceding collision, the scattering angle ϑ_i in the ith collision, and the azimuthal angle φ_i corresponding to it.

Now, we shall calculate the quantity $\cos\Theta_i$ averaged over a large number of particles passing through the material:

$$\langle\cos\Theta_i\rangle = \langle\cos\Theta_{i-1}\cos\vartheta_i\rangle$$
$$+\langle\sin\Theta_{i-1}\sin\vartheta_i\cos\varphi_i\rangle. \qquad (1.18)$$

Since, according to the hypothesis of independent successive collisions each collision takes place independently of the preceding one, the medium through which the particle passes is not ordered (i. e., it exhibits no particular directions), the distribution of collisions over the azimuthal angle φ_i is uniform. This means that $\langle\cos\varphi_i\rangle = 0$, i. e., the second term in (1.18) approaches zero, when averaging is performed over a large number of events. A consequence of the same hypothesis of independent pair collisions is that the mean product of the cosines in (1.18) can be replaced by the product of the corresponding mean quantities. As a result we obtain, upon averaging,

$$\langle\cos\Theta_i\rangle = \langle\cos\Theta_{i-1}\rangle\langle\cos\vartheta_i\rangle. \qquad (1.19)$$

Further, we shall consider only the case of fast heavy particles, in which multiple collisions of the particle with atoms of the medium result in the mean deviation angle being small. Then, taking advantage of the known expansion $\cos x \approx 1 - \frac{1}{2}x^2 + \ldots$, we obtain, instead of (1.19), the following:

$$\langle\Theta_i^2\rangle = \langle\Theta_{i-1}^2\rangle + \langle\vartheta_i\rangle^2. \qquad (1.20)$$

This recurrent relation shows that the mean square of the particle's deviation angle increases linearly with the average number of collisions:

$$\langle\Theta^2\rangle = N\langle\vartheta^2\rangle. \qquad (1.21)$$

Now, let us once again turn to Fig. 1.2 and formula (1.13). The average number of all collisions experienced by the particle upon passage through a thin layer of matter of thickness x is

$$N = n_0\sigma x, \qquad (1.22)$$

where σ is the total effective cross section of its interaction with an atom of the material. Before substituting this expression into $\langle \Theta^2 \rangle$, we shall consider the second factor in (1.21) and express the mean square scattering angle $\langle \vartheta^2 \rangle$ of the particle scattered from atom A in terms of the differential cross section $d\sigma/d\Omega$, which reflects the angular distribution of particles a due to the scattering process:

$$\langle \vartheta^2 \rangle = \int \vartheta^2 \frac{d\sigma}{d\Omega} d\Omega \Big/ \int \frac{d\sigma}{d\Omega} d\Omega \,. \tag{1.23}$$

The quantity $d\sigma/d\Omega$ present here gives the probability of particle 'a' undergoing scattering at an angle ϑ independently of whether the actual collision event turns out to be elastic or inelastic. In other words, the quantity $d\sigma/d\Omega$ occurring in (1.23) is the sum of differential cross sections of elastic and inelastic scattering:

$$\frac{d\sigma}{d\Omega} = \left(\frac{d\sigma}{d\Omega} \right)_{el} + \left(\frac{d\sigma}{d\Omega} \right)_{inel} \,. \tag{1.24}$$

Here [see (1.8)]

$$\int \frac{d\sigma}{d\Omega} d\Omega = \sigma \,, \qquad \int \left(\frac{d\sigma}{d\Omega} \right)_{el} d\Omega = \sigma_{el} \,, \tag{1.25}$$

$$\int \left(\frac{d\sigma}{d\Omega} \right)_{inel} d\Omega = \sigma_{inel} \,.$$

Thus, by substitution of (1.22) and (1.23) into (1.21) we obtain

$$\langle \Theta^2 \rangle = n_0 x \int \vartheta^2 \frac{d\sigma}{d\Omega} d\Omega \,, \tag{1.26}$$

the mean square deviation angle of a particle (the divergence of the particle beam) increases proportionally with the thickness of the layer of material covered. Relation (1.26) is the main formula of multiple scattering theory for particles passing through matter. Often, the quantity $\Theta = \sqrt{\langle \Theta^2 \rangle}$ is used; it is called the *mean angle of multiple scattering*.

In deriving formulae (1.16) and (1.26) assumptions were made restricting us to considering fast particles [see the arguments presented in obtaining relations (1.12) or (1.20)]. We now have to decide what particles can be considered "fast". Evidently, it is necessary that the velocity or kinetic energy of a particle passing through a medium is much greater than some quantity of the corresponding dimensionality characterizing either the medium itself or the interaction of our particle with an atom of the medium.

We shall apply two inequalities.

(a) The velocity of particle a is much greater than the mean speed of electrons in an atom of the medium:

$$v_a \gg \langle v_e \rangle \,. \tag{1.27a}$$

(b) The kinetic energy of particle a is much greater than the average potential energy of an interaction within the system a + A:

$$E_a \gg |\langle V_{\text{inter}} \rangle| . \tag{1.27b}$$

Let us estimate what this means in figures, say, in the case of protons interacting with matter.

From atomic physics we know that the mean energy of an electron on the nl orbit of a hydrogen atom or of a hydrogen-like ion is expressed through the Bohr velocity $v_B = \alpha c \approx 2 \times 10^8 \, \text{cm/s}$ by the formula

$$\langle v_e \rangle_{nl} = v_B \frac{Z}{n} , \tag{1.28}$$

where n is the principal quantum number, Z is the charge of the nucleus, and $\alpha = 1/137$ is the fine structure constant. If applied in the case of a multi-electron atom, formula (1.28) can be used only for approximate estimates by substituting the effective charge of the corresponding electron shell Z_{nl}^{eff} for Z; the mean velocity of an electron in an atom depends on the shell it occupies: the farther the electron shell considered is from the nucleus, the lower is the energy of the electron. Often, the Thomas–Fermi model is used for estimation of the electron energy averaged over all the shells of the atom:

$$\langle v_e \rangle_{\text{T-F}} \approx 0.7 v_B Z^{2/3} . \tag{1.29}$$

Consider the example of a proton passing through gaseous helium. In this case, the mean velocity of the electrons in the atom amounts to approximately $0.01 \, c$, and calculations readily reveal that the inequality $E_p \gg 50 \, \text{keV}$ corresponds to conditions (1.27). Thus, according to condition (1.27a), protons having kinetic energies of several MeV may surely be considered fast in a helium medium, so the above results can be applied in such a case. Interestingly, condition (1.27b) yields approximately the same figures as condition (1.27a).

1.2 The Classical Theory of Ionization Stopping

Now, let us combine conditions (1.27b) and (1.27a). On the one hand, the interaction of a fast particle 'a' with each atom of the medium is small, which means that the energy loss of the particle in each individual collision is also relatively small and, consequently, the curvature of its trajectory is negligible: in practice, the particle moves along a straight line. On the other hand, in accordance with the same conditions, the particles passes by an atom so rapidly, that it "sees" each of the electrons in the atom as if they were "frozen" at a certain point in space. The classic theory of ionization stopping of fast particles is based on the idea that a particle passing through a medium interacts with the atomic electrons as it does with independent free particles; here the energy losses of the particle due to its interaction with

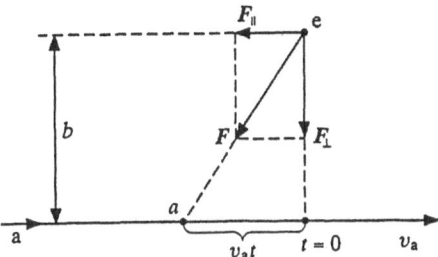

Fig. 1.4. Longitudinal and transverse components of the force F exerted on an electron by a passing particle

individual electrons just add up under the assumption that the electrons in the medium are distributed randomly (uniformly).

Now, consider the interaction of a fast, heavy charged particle with a single free electron. Let the particle pass by the electron with a constant speed v_a along a straight-line trajectory and with an impact parameter b (Fig. 1.4). We shall decompose the force $F(t)$, with which the particle a acts at each moment of time t on the electron, into longitudinal and transverse components, $F_\parallel(t)$ and $F_\perp(t)$, respectively. According to the conditions of the problem, formulated above, the longitudinal momentum $\int F_\parallel(t)\mathrm{d}t$ due to the force exerted, when the time $t < 0$, is fully compensated for by the momentum achieved when $t > 0$. At the same time, the transverse momentum due to the force turns out to be

$$\int F_\perp(t)\,\mathrm{d}t = 2\int_0^\infty F(t)\,\frac{b}{\sqrt{b^2 + (v_a t)^2}}\,\mathrm{d}t$$

$$= 2\int_0^\infty \frac{Z_a e^2}{b^2 + (v_a t)^2}\,\frac{b\,\mathrm{d}t}{\sqrt{b^2 + (v_a t)^2}} = \frac{2 Z_a e^2}{b v_a}. \tag{1.30}$$

We shall equate it to the momentum of the electron and calculate the electron's kinetic energy after its interaction with the particle that passed by:

$$T_e = \frac{p_e^2}{2m_e} = \frac{2 Z_a^2 e^4}{b^2 m_e v_a^2}. \tag{1.31}$$

Such is the loss of energy for a particle interacting with a single electron situated at a distance b from the particle's trajectory.

Now, consider a particle passing through a layer Δx in the medium. Imagine a thin annular cylinder cut out of it, so that its axis coincides with the trajectory of motion of the particle (Fig. 1.5). The number of electrons inside this annular cylinder equals

$$\mathrm{d}N_e = n_e 2\pi b\,\mathrm{d}b\,\Delta x\,, \tag{1.32}$$

where n_e is the density of electrons in the medium, and the energy loss of the particle due to its interaction with these electrons amounts to $T_e \mathrm{d}N_e$, where T_e is determined by formula (1.31). To obtain the total energy loss of

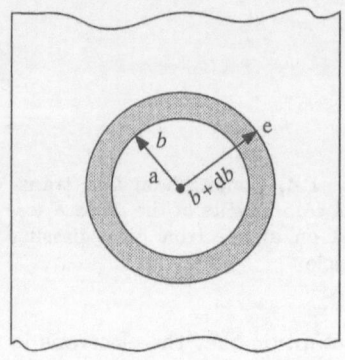

Fig. 1.5. Cross section of an annular cylinder $(b, b + \mathrm{d}b)$

a particle passing through a layer Δx it remains to perform integration over the entire plane:

$$-\Delta E = \int T_\mathrm{e}\mathrm{d}N_\mathrm{e} = n_\mathrm{e}\Delta x \int\limits_0^\infty T_\mathrm{e}2\pi b\,\mathrm{d}b. \qquad (1.33)$$

Hence, for the stopping power of a medium we obtain

$$-\frac{\mathrm{d}E}{\mathrm{d}x} = \frac{4\pi Z_\mathrm{a}^2 e^4}{m_\mathrm{e}v_\mathrm{a}^2}\,n_\mathrm{e}\int\frac{\mathrm{d}b}{b}. \qquad (1.34)$$

Here, we shall interrupt our calculations and discuss what must be done with the integral. If we adopt a formal approach, this integral will diverge. From a physical point of view, however, it would be clearly unreasonable to perform integration from zero up to infinity, since at both extremely small and extremely large b values we are beyond the reach of the assumptions underlying relation (1.34). Indeed, when $b \to 0$, there is no sense in considering the electron to retain its postion in space after its interaction with the particle: the electron acquires a large energy in such a collision. On the other hand, in the case of very large b the energy received by the electron turns out to be lower than the ionization potential and even lower than the excitation threshold of the atom; in this case the assumption that the electrons are free is in sharp contrast with the actual stucture of the medium. Therefore, we shall restrict the integration in (1.34) to finite limits, from some b_min up to a certain b_max:

$$-\frac{\mathrm{d}E}{\mathrm{d}x} = \frac{4\pi Z_\mathrm{a}^2 e^4}{m_\mathrm{e}v_\mathrm{a}^2}\,n_\mathrm{e}\ln\frac{b_\mathrm{max}}{b_\mathrm{min}}. \qquad (1.35)$$

The above physical arguments can be used to derive the following formula for the logarithmic factor present in (1.35):

$$\ln\frac{b_\mathrm{max}}{b_\mathrm{min}} \approx \ln\frac{2m_\mathrm{e}v_\mathrm{a}^2}{I}, \qquad (1.36)$$

where I is a quantity of the order of magnitude of the mean excitation energy or of the ionization potential of the atom. However, we shall not perform the relevant derivation now, since later, in presenting a more consistent quantum approach, we shall obtain a formula similar to (1.35) with its own logarithmic factor and estimate that. Meanwhile, we shall note only that the logarithmic factor in (1.35) is quite a *flasque* function of the velocity (kinetic energy) of the passing particle, so the main properties of the stopping power of a medium $(-dE/dx)$ are determined by the prelogarithmic factor.

Let us analyze formula (1.35). First, we see that the dependence of ionization losses upon the kinetic energy E and the particle mass m_a reduces entirely to the dependence of their ratio on E, i. e., the particle velocity:

$$-\frac{dE}{dx} \sim \frac{1}{v_a^2} \sim \frac{m_a}{E} \, . \tag{1.37}$$

Second, the ionization losses are proportional to the square of the charge of the particle passing through the medium:

$$-\frac{dE}{dx} \sim Z_a^2 \, . \tag{1.38}$$

This means, for instance, that the stopping powers of a medium will differ [within the range of applicability of (1.35)] by a factor of 4 in the case of protons and α particles with identical velocities, whereas in the case of identical kinetic energies the stopping powers will difffer by a factor of 16:

$$\frac{\left(-\frac{dE}{dx}\right)_p}{\left(-\frac{dE}{dx}\right)_\alpha}\bigg|_{v_p=v_\alpha} \approx \frac{1}{4} \, ; \qquad \frac{\left(-\frac{dE}{dx}\right)_p}{\left(-\frac{dE}{dx}\right)_\alpha}\bigg|_{E_p=E_\alpha} \approx \frac{1}{16} \, . \tag{1.39}$$

Formula (1.35) shows that the ionization losses are proportional to the density of electrons in the medium. As done in Sect. 1.1, we denote the density of atoms in the medium by n_0. Then, if Z is the atomic number of the medium,

$$n_e = Z n_0 \, . \tag{1.40}$$

Consequently, (1.35) can be rewritten as

$$-\frac{dE}{dx} = \frac{4\pi Z_a^2 e^4}{m_e v_a^2} Z n_0 \ln \frac{b_{max}}{b_{min}} \, . \tag{1.41}$$

If we take into account that the atomic density n_0 is related to the density of the medium, ρ, via the Avogadro number, $N_0 = 6.022 \times 10^{23} \, \mathrm{mol}^{-1}$, and that the mass number of the atoms of the medium, A, is

$$n_0 = \rho\left(\frac{N_0}{A}\right) , \tag{1.42}$$

(1.41) is transformed further into

$$-\frac{dE}{dx} = \frac{4\pi Z_a^2 e^4}{m_e v_a^2} \frac{Z}{A} \rho N_0 \ln \frac{b_{max}}{b_{min}} \, . \tag{1.43}$$

Taking into account that the ratio Z/A varies from substance to substance within quite a narrow range (for $^4_2$He it is 0.5; for $^{208}_{82}$Pb it is 0.4), we see that the stopping power of a substance is, roughly speaking, proportional to its density:

$$-\frac{\mathrm{d}E}{\mathrm{d}x} \sim \rho. \tag{1.44}$$

Besides the stopping power, we have also introduced, in Sect. 1.1, the so-called mass stopping power $-\mathrm{d}E/\mathrm{d}X = \frac{1}{\rho}(\mathrm{d}E/\mathrm{d}x)$. We now see that it varies very slightly from substance to substance:

$$-\frac{\mathrm{d}E}{\mathrm{d}X} = \frac{4\pi Z_a^2 e^4}{m_e v_a^2} \frac{Z}{A} N_0 \ln \frac{b_{\max}}{b_{\min}}; \tag{1.45}$$

in other words, the energy losses of particles passing through layers of equal thickness, expressed in $\mathrm{g/cm}^2$, are approximately the same in different substances.

Experimental data on the stopping power of lead are presented in Fig. 1.6. We see that its dependence on the particle velocity is much more complicated than can be expected from the simple classical theory presented above.

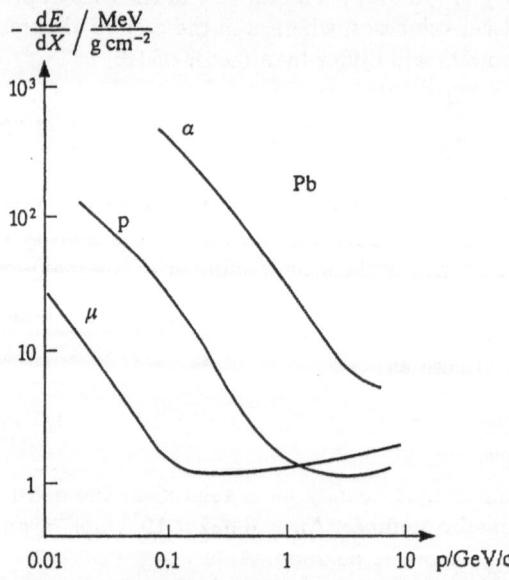

Fig. 1.6. Stopping power of lead for protons, muons, and α particles

Second Lecture

2.1 Quantum Theory of Ionization Stopping of Fast Charged Particles

At the beginning of the preceding lecture we expressed the parameters characterizing the passage of charged particles through matter with the aid of the characteristics of elementary processes of interaction between a particle and individual atoms in the medium. We then set aside these relations and, turning to classical theory of ionization stopping, applied such significant simplifications that the discrete, atomic structure of matter remained beyond the scope of our examination. We shall now continue the consistent analysis initatiated in Sect. 1.1, which resulted in (1.16) and (1.27). Our task will now consist in finding, and substituting into these formulae, the characteristics of the particle's interaction with the atoms of the medium: the effective stopping $S(E)$ and the differential scattering cross section $d\sigma/d\Omega$. To this end we turn to quantum collision theory.

In quantum theory, the differential cross sections of elastic scattering, in which the atom remains in its initial state, or of inelastic scattering, which is accompanied by $|0\rangle \rightarrow |n\rangle$ transitions from one atomic state to another, are expressed through the corresponding scattering amplitude:

$$\left(\frac{d\sigma}{d\Omega}\right)_{el} = |F_{el}(E,\vartheta)|^2 , \qquad (2.1)$$

$$\left(\frac{d\sigma_n}{d\Omega}\right) = \frac{k'}{k}\left|F_{0\rightarrow n}(E,\vartheta)\right|^2 , \qquad (2.2)$$

where E is the energy of the incident particle and ϑ is the scattering angle. For determining the scattering amplitude it is generally necessary to solve a certain wave equation (or even a set of coupled equations) and to extract these amplitudes from the asymptotes of the corresponding wave functions. However, if the kinetic energy of the incident particle is much greater than its average energy of interaction with the atoms (and, in accordance with (1.27b), this is precisely the case of the "fast particles" we are interested in), then the Born approximation can be used, within the framework of which the problem is essentially simplified.

The Born amplitude of elastic scattering $a + A \rightarrow a + A$ is simply the integral

$$F_{\mathrm{el}}^{(\mathrm{Born})}(E, \vartheta) = -\frac{m_{\mathrm{a}}}{2\pi\hbar^2} \int e^{i\mathbf{k}' \cdot \mathbf{r}} V(\mathbf{r}) e^{i\mathbf{k} \cdot \mathbf{r}} d^3 r \,, \tag{2.3}$$

which contains the plane wave $\varphi_k = e^{i\mathbf{k} \cdot \mathbf{r}}$ describing free motion of particle a in the initial state, the wave function of the particle in the final state, $\varphi_{k'} = e^{i\mathbf{k} \cdot \mathbf{r}}$, and the interaction potential of the atom with the particle averaged over its ground state $|0\rangle$:

$$V(\mathbf{r}) = \frac{Z_{\mathrm{a}} Z e^2}{r} - \left\langle 0 \left| \sum_{j=1} \frac{Z_{\mathrm{a}} e^2}{|\mathbf{r} - \mathbf{r}_j|} \right| 0 \right\rangle \tag{2.4}$$

[to be precise, instead of the mass of the incident particle, it is the reduced mass of the $a + A$ system, $\mu = m_{\mathrm{a}} - m_{\mathrm{A}}/m_{\mathrm{a}} + m_{\mathrm{A}}$, that actually occurs in (2.3)]. The wave vectors $\mathbf{k}$ and $\mathbf{k}'$ in (2.3) represent, with an accuracy up to Planck's constant, the momenta of the incident and scattered particles, respectively: $\mathbf{k} = \mathbf{p}/\hbar$; $\mathbf{k}' = \mathbf{p}'/\hbar$; below we shall just call vector $\mathbf{k}$ or $\mathbf{k}'$ the particle momentum.

From (2.3) it is seen that in the Born approximation the elastic scattering amplitude is the sum of two terms: the scattering amplitude due to interaction of particle a with the atomic nucleus and the scattering amplitude due to interaction with the atomic electrons; the whole scattering pattern depends on the interference of these two amplitudes. The first summand is calculated by straightforward integration:

$$F_{\mathrm{el,nucl}}^{(\mathrm{Born})}(E, \vartheta) = -\frac{\mu}{2\pi\hbar^2} \int \frac{Z_{\mathrm{a}} Z e^2}{r} e^{i(\mathbf{k} - \mathbf{k}')\mathbf{r}} d^3 r = \frac{2\mu Z_{\mathrm{a}} Z e^2}{\hbar^2 q^2} \,. \tag{2.5}$$

Here, we have introduced the momentum transfer vector

$$\mathbf{q} = \mathbf{k} - \mathbf{k}' \,; \tag{2.6}$$

this is the only kinematic variable upon which the Born scattering amplitude depends. The dependence of the value and orientation of the vector $\mathbf{q}$ on the scattering angle of the particle is clear from Fig. 2.1.

For calculating the second term in the elastic scattering amplitude, we shall introduce the electron density

$$\rho_{\mathrm{elec}}(\mathbf{r}) = \left\langle 0 \left| \sum_{j=1}^{Z} \delta(\mathbf{r} - \mathbf{r}_j) \right| 0 \right\rangle \,, \tag{2.7}$$

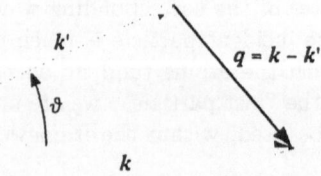

Fig. 2.1. Kinematics of elastic and inelastic scattering of a particle

in terms of which we shall express the average interaction potential of the particle with the electron shell:

$$\left\langle 0 \left| \sum_{j=1}^{Z} \frac{Z_a e^2}{|r - r_j|} \right| 0 \right\rangle = \int \frac{Z_a e^2}{|r - r'|} \rho_{\text{elec}}(r') \mathrm{d}^3 r' . \tag{2.8}$$

Substitution of (2.8) into (2.4) and then into (2.3), and use of (2.6) give

$$F_{\text{el,elec}}^{(\text{Born})}(E, \vartheta) = \frac{2\mu Z_a e^2}{\hbar^2 q^2} \int \rho_{\text{elec}}(r) e^{i q \cdot r} \mathrm{d}^3 r . \tag{2.9}$$

At small q the integral present here can be readily seen to tend toward Z. At $q \gg 1/a$, where a is of the same order of magnitude as the size of the atom, it decays rapidly owing to oscillations of the exponential factor $e^{i q \cdot r}$. In the $0 < q < 1/a$ region, the dependence of this integral on q is determined by the shape of the electron density distribution, $\rho_{\text{elec}}(r)$. Let us now define the *electron density form factor* of the atom,

$$\mathcal{F}_{\text{elec}}(q) = \frac{1}{Z} \int \rho_{\text{elec}}(r) e^{i q \cdot r} \mathrm{d}^3 r = \frac{1}{Z} \left\langle 0 \left| \sum_{j=1}^{Z} e^{i q \cdot r_j} \right| 0 \right\rangle , \tag{2.10}$$

which, up to the factor $1/Z$, is the Fourier transform of the electron density. Upon collecting the summands in (2.5) and (2.9), we express the elastic scattering amplitude of the electron on the atom through $\mathcal{F}_{\text{elec}}(q)$

$$F_{\text{el}}(E, \vartheta) = -\frac{2\mu Z_a Z e^2}{\hbar^2 q^2} \left[1 - \mathcal{F}_{\text{elec}}(q) \right] . \tag{2.11}$$

Here and below, we shall drop the index "Born", although we shall remember that this expression is not exact, but has been obtained in the Born approximation.

Thus, the differential elastic cross section has the form

$$\left(\frac{\mathrm{d}\sigma}{\mathrm{d}\Omega} \right)_{\text{el}} = \frac{4\mu^2 Z_a^2 Z^2 e^4}{\hbar^4 q^4} \left| 1 - \mathcal{F}_{\text{elec}}(q) \right|^2 . \tag{2.12}$$

When $q \gg 1$, it transforms into the scattering cross section of a particle of charge Z_a on a point-like center of charge Z and is described by Rutherford's formula

$$\left(\frac{\mathrm{d}\sigma}{\mathrm{d}\Omega} \right)_{\text{el}} \bigg|_{q \gg 1/a} \to \left(\frac{\mathrm{d}\sigma}{\mathrm{d}\Omega} \right)_{\text{R}} . \tag{2.13}$$

We shall use $(\mathrm{d}\sigma/\mathrm{d}\Omega)_{\text{R}}$ for the Rutherford cross section corresponding to the scattering of a particle of charge Z_a from a unit point-like charge ($Z = 1$):

$$\left(\frac{\mathrm{d}\sigma}{\mathrm{d}\Omega} \right)_{\text{R}} \equiv \frac{4\mu^2 Z_a^2 e^4}{\hbar^4 q^4} = \frac{Z_a^2 e^4}{16 E^2 \sin^4 \frac{\vartheta}{2}} . \tag{2.14}$$

In this notation the differential elastic scattering cross section of particle a scattering from the atom assumes the form

$$\left(\frac{d\sigma}{d\Omega}\right)_{el} = Z^2 \left(\frac{d\sigma}{d\Omega}\right)_R \left|1 - \mathcal{F}_{elec}(q)\right|^2. \tag{2.15}$$

From the formula obtained it can be seen that at large momentum transfers $q \gg 1/a$, i.e., when the particle undergoes large-angle deviations, the scattered particle does not feel the electron shell of the atom and undergoes scattering from the bare nucleus. On the contrary, when $q \ll 1/a$ (extremely small scattering angles), the electron shell screens the nucleus, and the scattering cross section turns out to be essentially smaller than the Rutherford cross section.

In the Born approximation, the cross section of inelastic scattering $a + A \to a' + A_n^*$ is calculated as above for the case of elastic scattering and is expressed for each partial transition $|0\rangle \to |n\rangle$ in terms of the so-called *transition, or inelastic, form factor* $\mathcal{F}_{0 \to n}(q)$:

$$\frac{d\sigma_n}{d\Omega} = \frac{k'}{k} \left(\frac{d\sigma}{d\Omega}\right)_R \left|\mathcal{F}_{0 \to n}(q)\right|^2, \tag{2.16}$$

$$\mathcal{F}_{0 \to n}(q) = \left\langle n \left| \sum_{j=1}^{Z} e^{iq' \cdot r_j} \right| 0 \right\rangle. \tag{2.17}$$

Note that if the energy of the incident particles is much higher than the excitation energy of the atom, the kinematic factor k'/k occurring in (2.16) practically coincides with unity:

$$\frac{k'}{k} = \sqrt{1 - \frac{\varepsilon_n - \varepsilon_0}{E}} \to 1. \tag{2.18}$$

We have thus prepared all the necessary intermediate products from quantum collision theory necessary for our quantum-mechanical examination of the ionization stopping of fast charged particles. Let us proceed to calculate the effective stopping

$$S(E) = \sum_{n} (\varepsilon_n - \varepsilon_0)\sigma_n \tag{2.19}$$

for which we shall first calculate the integral partial cross sections of inelastic scattering,

$$\sigma_n = \int \frac{d\sigma_n}{d\Omega} \, d\Omega. \tag{2.20}$$

Since when $k'/k \to 1$ the differential cross section $d\sigma_n/d\Omega$ depends on only one kinematic variable q, it is convenient to pass in (2.20) from integration over the scattering angle ϑ to integration over q. When the scattering angle ϑ varies from zero up to π, the momentum transfer varies from $q_{min} = k - k'$ up to $q_{max} = k + k'$ in accordance with the law (see Fig. 2.1)

$$q^2 = k^2 + k'^2 - 2kk' \cos\vartheta. \tag{2.21}$$

We shall take into account the law of energy conservation,

$$E - E' = \varepsilon_n - \varepsilon_0 \,, \tag{2.22}$$

as well as the fact that the excitation energy is negligible as compared with the energy of the incident particles,

$$E - E' \ll E \,. \tag{2.23}$$

Then, with a good precision, we shall have

$$q_{\min} = \frac{k^2 - k'^2}{k + k'} \approx \frac{\varepsilon_n - \varepsilon_0}{\hbar v_a} \,; \quad q_{\max} \approx 2k \,. \tag{2.24}$$

Moreover, taking into account (2.21), we have

$$d\Omega = 2\pi \sin \vartheta \, d\vartheta = 2\pi \frac{q \, dq}{kk'} \approx \frac{2\pi q \, dq}{k^2} \,. \tag{2.25}$$

Thus, the partial cross sections can be calculated by the formula

$$\sigma_n = \frac{2\pi}{k^2} \int_{q_{\min}}^{q_{\max}} \frac{d\sigma_n}{d\Omega} q \, dq \,. \tag{2.26}$$

Now we substitute this expression into (2.19) and take $d\sigma_n/d\Omega$ from (2.16) $(k'/k \to 1)$:

$$\begin{aligned}
S(E) &= \frac{2\pi}{k^2} \sum_n (\varepsilon_n - \varepsilon_0) \int_{q_{\min}}^{q_{\max}} \frac{d\sigma_n}{d\Omega} q \, dq \\
&= \frac{2\pi}{k^2} \frac{4\mu^2 Z_a^2 e^4}{\hbar^4} \sum_n (\varepsilon_n - \varepsilon_0) \int_{q_{\min}}^{q_{\max}} \left| \left\langle n \left| \sum_{j=1}^{Z} e^{i q \cdot r_j} \right| 0 \right\rangle \right|^2 \frac{q \, dq}{q^4} \,.
\end{aligned} \tag{2.27}$$

In quantum mechanics, the validity of the following formula is proven:

$$\sum_n (\varepsilon_n - \varepsilon_0) \left| \left\langle n \left| \sum_{j=1}^{Z} e^{i q \cdot r_j} \right| 0 \right\rangle \right|^2 = Z \frac{\hbar^2 q^2}{2m_e} \,, \tag{2.28}$$

with the aid of which the infinite sum over all the excited states of the atom is reduced exactly to a simple analytical expression. Regretfully, we cannot apply it directly to (2.27), since the lower integration limit $q_{\min}$ depends on n itself [see (2.24)]. Therefore, we shall replace the lower limit in (2.27) by its mean value:

$$q_{\min} \to \bar{q}_{\min} = \frac{\overline{(\varepsilon_n - \varepsilon_0)}}{\hbar v_a} \equiv \frac{I}{\hbar v_a} \,. \tag{2.29}$$

We may now transpose the orders of summation and integration in (2.27). As a result we obtain

$$S(E) = \frac{2\pi}{k^2} \frac{4\mu^2 Z_a^2 e^4}{\hbar^4} \int\limits_{q_{min}}^{q_{max}} \frac{dq}{q^3} \sum_n (\varepsilon_n - \varepsilon_0) \left| \left\langle n \left| \sum_{j=1}^{Z} e^{i q \cdot r_j} \right| 0 \right\rangle \right|^2$$

$$= \frac{4\pi Z_a^2 e^4 Z}{m_e v_a^2} \ln \frac{q_{max} \hbar v_a}{I}. \tag{2.30}$$

One can see here the parameter I, which is the mean excitation energy of the atom, brought about by the passing particle. A consistent calculation of this quantity will be performed in Sect. 3.2, and therein the following formula will also be obtained, which expresses I through the transition oscillator strengths f_{n0}:

$$\ln I = \frac{1}{Z} \sum_n f_{n0} \ln(\varepsilon_n - \varepsilon_0). \tag{2.31}$$

Application of this formula in the case of hydrogen, for example, yields $I = 14.9\,\mathrm{eV}$. This exceeds the ionization potential $I_{\mathrm{ioniz}} = 13.6\,\mathrm{eV}$. Thus, the main contribution to the stopping of a particle is indeed, as noted in Sect. 1.1, due to ionization of the atom, and not due to excitation of its discrete levels.

Now, let us again turn to the above expression (2.30) for effective stopping, in which we have not yet revealed the meaning of the symbol q_{max}. Formally, $q_{max} = 2k$, and this value is attained when the incident particle scatters from the atom backwards. However, this may happen very rarely and even then, only when our heavy incident particle interacts with a particle that is also heavy, i.e. with the nucleus of an atom. But at present we are interested in the excitation or ionization process of the atom, when energy is transferred directly from the incident particle to the atomic electron. Therefore, q_{max} in (2.30) is actually determined by the collision kinematics of the collision between a heavy particle and an electron.

If this electron were at rest before the collision, then from the kinematics of an elastic collision, $a+e \rightarrow a+e$, it would be easy to calculate the maximum momentum of the electron after the collision: with high precision, it is $2m_e v_a$, where v_a is the velocity of the incident particle. In our conditions [remember relation (1.27)] the quantity $2m_e v_a$ is significantly greater than the average momentum of an electron in the atom before it undergoes interaction with the heavy particle. Therefore the motion of the electron in the atom can be neglected, and the value obtained may be used in estimating the maximum momentum transferred by the incident particle to the atomic electron:

$$\hbar q_{max} \approx 2m_e v_a. \tag{2.32}$$

It now remains for us to substitute (2.32) into (2.30):

$$S(E) = \frac{4\pi Z_a^2 e^4 Z}{m_e v_a^2} \ln \frac{2m_e v_a^2}{I}. \tag{2.33}$$

We now multiply this expression by the concentration of atoms in the medium and obtain the stopping power of the medium:

$$-\frac{\mathrm{d}E}{\mathrm{d}x} = \underbrace{\frac{4\pi Z_a^2 e^4}{m_e v_a^2}}_{\substack{\text{main depen-}\\ \text{dence on}\\ \text{particle}\\ \text{velocity}}} \underbrace{Z n_0}_{\substack{\text{main dependence}\\ \text{on properties of}\\ \text{the medium}}} \underbrace{\ln \frac{2 m_e v_a^2}{I}}_{\substack{\text{weak depen-}\\ \text{dence on par-}\\ \text{ticle velocity}\\ \text{and properties}\\ \text{of the medium}}} \qquad (2.34)$$

This final formula, obtained within the framework of the quantum approach, exhibits the same principal features as (1.41). It may be considered as a justification of the classical approach applied in deriving formula (1.41). At the same time, we have pushed forward, somewhat, by removing the uncertainties concerning the logarithmic factor that remained in Sect. 2.1 and by revealing the physical meaning of the parameters occurring in it.

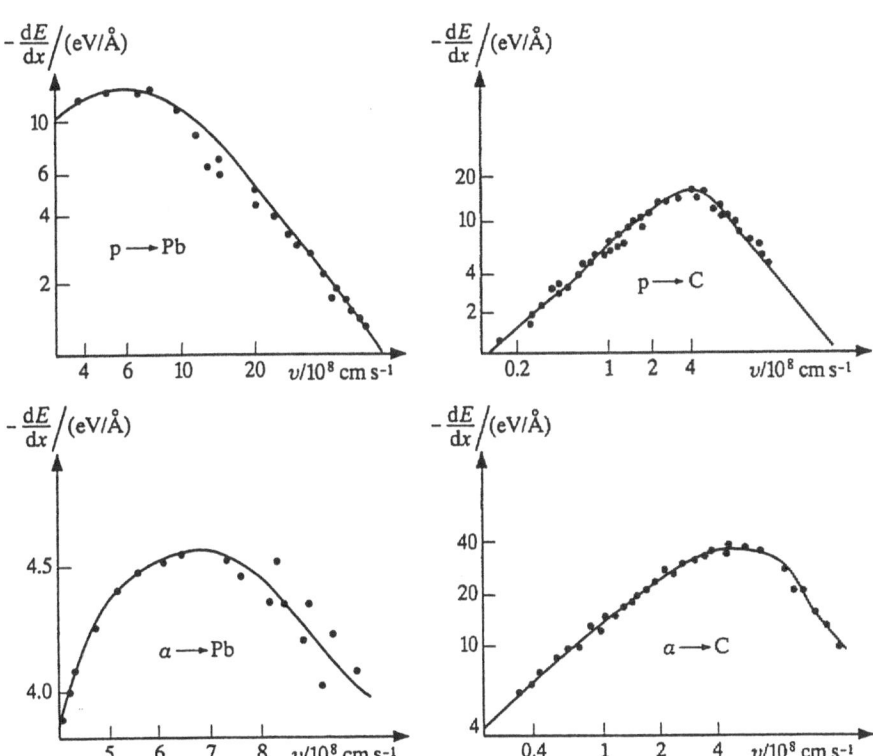

Fig. 2.2. Stopping power of graphite and lead for protons and α particles [2.1]

Figure 2.2 presents the stopping powers of graphite and lead for protons and α particles. It is clearly seen that the experimentally established dependence of the stopping power on the particle velocity (energy), if considered throughout the entire range of these variables, is significantly more complicated than what follows from our formula (2.34). In the next lecture we shall discuss various corrections to this dependence, but for the time being we shall note only that for a significant part of the $-\mathrm{d}E/\mathrm{d}x$ curve, corresponding to proton energies of approximately hundreds of keV up to a hundred MeV, the simple law $-\mathrm{d}E/\mathrm{d}x \sim 1/v_\mathrm{a}^2$ is obeyed, and for this reason we shall take advantage of it for obtaining qualitative estimates.

2.2 Stopping Times and Path Ranges of Particles in Media

A fast particle passing through a medium spends its energy on excitation and ionization of atoms of the medium until it finally "stops", i. e., starts taking part in the thermal motion of the atoms of the medium. How soon does this happen?

Let us now turn to formula (2.34). Taking into account that $\mathrm{d}E = m_\mathrm{a}v_\mathrm{a}\mathrm{d}v_\mathrm{a}$ and $\mathrm{d}x = v_\mathrm{a}\mathrm{d}t$ we write it in an equivalent form:

$$-\frac{\mathrm{d}v_\mathrm{a}}{\mathrm{d}t} = \frac{4\pi Z_\mathrm{a}^2 e^4}{m_\mathrm{a}m_e v_\mathrm{a}^2} Z n_0 \ln \frac{2m_e v_\mathrm{a}^2}{I} . \tag{2.35}$$

From the following we obtain for the time the particle takes to slow from the initial velocity v_in down to a certain velocity v:

$$t(v_\mathrm{in} \rightarrow v) = \int\limits_{v_\mathrm{in}}^{v} -\frac{m_\mathrm{a}m_e v_\mathrm{a}^2}{4\pi Z_\mathrm{a}^2 e^4 Z n_0 \ln(2m_e v_\mathrm{a}^2/I)} \mathrm{d}v_\mathrm{a} . \tag{2.36}$$

This time is proportional to the mass m_a and inversely proportional to the square of the particle charge Z_a. If we neglect the fact that the velocity of the particle is present not only in the integrand as v_a^2, but also in the logarithm, then the integral (2.36) is readily computed, and the dependence of the stopping time on the initial velocity is revealed: $t \sim v_\mathrm{in}^3 - v^3$. The result obtained is quite consistent with what (2.34) yields for the stopping power: namely the deceleration process is the faster the lower the velocity of the particle. In this connection we note that the last stages of the stopping process are so rapid that their contribution to the total stopping time is negligible. Therefore, one can simplify the concept of the stopping time of a particle in a medium and calculate it as the time required for the velocity of a particle to drop from the initial value v_in down to zero. Then, in accordance with (2.36), we have

$$t(v_\mathrm{in}) \sim \frac{m_\mathrm{a}}{Z_\mathrm{a}^2 Z n_0} v_\mathrm{in}^3 \sim \frac{E^{3/2}}{\sqrt{m_\mathrm{a}} Z_\mathrm{a}^2 Z n_0} , \tag{2.37}$$

i. e., the stopping time of a particle is roughly proportional to its cubic velocity (the energy raised to the power 3/2). Naturally, these estimates are reasonable only when the initial velocity of the particle happens to correspond to the falling part of the $-dE/dx$ curve, and only to the extent to which the $1/v_a^2$ law holds valid.

Let us also obtain similar semiqualitative estimates for the particle path ranges:

$$R(E) = \int\limits_0^R dx = \int\limits_0^{v_{in}} \frac{dE'}{(-dE/dx)_{E'}} = \int\limits_0^{v_{in}} \frac{m_e v_a^2 m_a v_a dv_a}{4\pi Z_a^2 e^4 Z n_0 \ln(2m_e v_a^2/I)} . \quad (2.38)$$

Whatever the initial velocity v_{in}, the total path range of the particle, R, like its stopping time t, is proportional to the particle mass and inversely proportional to the square of its charge. At the $1/v_a^2$ part of the curve, the analytic dependence of the path on the initial particle velocity (initial energy) is revealed.

$$R \sim \frac{m_a v_a^4}{Z_a^2 Z n_0} \sim \frac{E^2}{m_a Z_a^2 Z n_0} . \quad (2.39)$$

Consider a small example. According to (2.39), the path ranges of protons and α particles with identical initial energies in one and the same medium should differ by a factor of 16. Let us compare this figure with what is obtained in experiments. In aluminium the path ranges of 10 MeV protons amount to 630 μm (0.6 mm), and the path ranges of α particles of the same energy are 61 μm; their ratio is not 16, but does not deviate from this figure so much.

A clear idea of the passage of charged particles through a medium is given by the so-called *Bragg curves*, which show the distribution of the specific ionization produced by a particle along the trajectory ("track") of its motion (Fig. 2.3). In accordance with (2.34), at the beginning of the track, before the particle has lost its velocity, the specific ionization is low and increases slowly and monotonously as the the stopping process of the particle proceeds. By the end of the path the specific ionization rises sharply and, finally, rapidly drops down to zero.

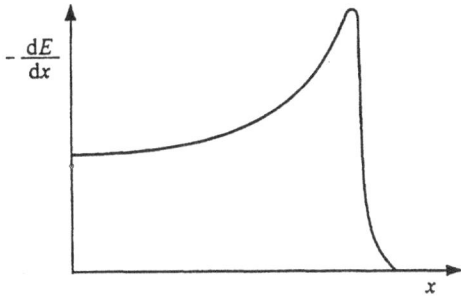

Fig. 2.3. Specific ionization due to a particle passing through a substance (the "Bragg curve")

2.3 Calculation
of the Average Multiple-Scattering Angle

In accordance with (1.21), knowledge of the differential scattering cross section of particles scattering from atoms of the medium is required for calculation of the average multiple scattering angle $\Theta = \sqrt{\langle\Theta^2\rangle}$. Elastic and inelastic scattering give additive contributions to $\langle\Theta^2\rangle$:

$$\langle\Theta^2\rangle = n_0 x \int \vartheta^2 \left[\left(\frac{d\sigma}{d\Omega}\right)_{\text{el}} + \left(\frac{d\sigma}{d\Omega}\right)_{\text{inel}}\right] d\Omega. \tag{2.40}$$

We shall clarify the relationship between these contributions.

Let us start with inelastic scattering. In the Born approximation, the differential cross section of inelastic scattering of a fast particle scattering from an atom is given by (2.16) and (2.17):

$$\left(\frac{d\sigma}{d\Omega}\right)_{\text{inel}} \approx \left(\frac{d\sigma}{d\Omega}\right)_R \sum_{n\neq0}\left|\left\langle n\left|\sum_{j=1}^{Z} e^{i\boldsymbol{q}\cdot\boldsymbol{r}_j}\right|0\right\rangle\right|^2 \tag{2.41}$$

(we have set $k'/k = 1$, here). The "sizes" of the atomic wave functions, for which the matrix element $\langle n|\sum e^{i\boldsymbol{q}\cdot\boldsymbol{r}j}|0\rangle$ is computed, are of the order of the atomic length a. This defines two limit regions for the scattering angles or, which is essentially the same, for the momentum transfer q:

$$q \ll \frac{1}{a}; \quad q \gg \frac{1}{a}. \tag{2.42}$$

In the first case, the exponent is small throughout the whole inner region of the atom, so advantage may be taken of the following expansion of the exponential function in a Taylor series in calculating the matrix elements:

$$\left\langle n\left|\sum_{j=1}^{z} e^{i\boldsymbol{q}\cdot\boldsymbol{r}_j}\right|0\right\rangle = \left\langle n\left|\sum_{j}(1 + i\boldsymbol{q}\cdot\boldsymbol{r}_j + \ldots)\right|0\right\rangle$$

$$= i\boldsymbol{q}\left\langle n\left|\sum_{j}\boldsymbol{r}_j\right|0\right\rangle = iq\left\langle n\left|\sum_{j}x_j\right|0\right\rangle. \tag{2.43}$$

And, further,

$$\sum_{n\neq1}\left|\left\langle n\left|\sum_{j} e^{i\boldsymbol{q}\cdot\boldsymbol{r}_j}\right|0\right\rangle\right|^2 \approx q^2 \sum_{n\neq0}\left|\left\langle n\left|\sum_{j}x_j\right|0\right\rangle\right|^2$$

$$= q^2\left\langle 0\left|\sum_{j,j'}x_j x_{j'}\right|0\right\rangle = q^2\left\langle 0\left|\sum_{j}x_j^2\right|0\right\rangle \sim Zq^2a^2. \tag{2.44}$$

Thus, for estimating the differential cross section of inelastic scattering in the region of small momentum transfers we obtain

$$\left(\frac{d\sigma}{d\Omega}\right)_{\text{inel}}\bigg|_{q\ll1/a} \approx \left(\frac{d\sigma}{d\Omega}\right)_R Zq^2a^2 \sim \frac{Z}{q^2}. \tag{2.45}$$

In passing to the region of large momentum transfers $(q \gg 1/a)$ we take advantage of the properties of completeness of atomic wave functions, (1.5):

$$\sum_{n \neq 0} \left| \left\langle n \left| \sum_j e^{iq \cdot r_j} \right| 0 \right\rangle \right|^2 = \left\langle 0 \left| \sum_{j,j'} e^{iq(r_j - r_{j'})} \right| 0 \right\rangle$$

$$- \left| \left\langle 0 \left| \sum_j e^{iq \cdot r_j} \right| 0 \right\rangle \right|^2 \rightarrow Z + \left\langle 0 \left| \sum_{j \neq j'} e^{iq(r_j - r_{j'})} \right| 0 \right\rangle^2 \rightarrow Z . \quad (2.46)$$

The crossed-out terms have been discarded because the exponentials in these terms oscillate very strongly when $q \gg 1/a$. Now, substituting (2.46) into (2.41) we find that at large momentum transfers the inelastic scattering of a fast particle from an atom is equivalent to its being scattered from Z free electrons:

$$\left(\frac{d\sigma}{d\Omega} \right)_{\text{inel}} \Big|_{q \gg 1/a} \approx Z \left(\frac{d\sigma}{d\Omega} \right)_R \sim \frac{Z}{q^4} . \quad (2.47)$$

Note that the differential cross section $(d\sigma/d\Omega)_{\text{inel}}$ is proportional to the atomic number Z in both limit cases, $q \ll 1/a$ and $q \gg 1/a$.

Now, we turn to elastic scattering:

$$\left(\frac{d\sigma}{d\Omega} \right)_{\text{el}} = Z^2 \left(\frac{d\sigma}{d\Omega} \right)_R \left| 1 - \mathcal{F}_{\text{elec}}(q) \right|^2 . \quad (2.48)$$

From the general formula (2.10) it is seen that at small momentum transfers the form factor $\mathcal{F}_{\text{elec}}(q)$ is expressed via the root-mean-square radius of the electron shell of the atom,

$$\mathcal{F}_{\text{elec}}(q) \approx 1 - \frac{1}{6} q^2 \langle r^2 \rangle + \dots . \quad (2.49)$$

Substitution of this expression into (2.48) yields

$$\left(\frac{d\sigma}{d\Omega} \right)_{\text{el}} \Big|_{q \ll 1/a} \approx \frac{Z^2}{36} \left(\frac{d\sigma}{d\Omega} \right)_R q^4 (\langle r^2 \rangle)^2 . \quad (2.50)$$

Taking into account that the Rutherford cross section $(d\sigma/d\Omega)_R$ is proportional to $1/q^4$, we see that as $q \rightarrow 0$, the cross section of elastic scattering of a fast charged particle from an atom of the medium tends toward a constant.

In Sect. 2.1, we revealed that when $q \gg 1/a$, the form factor of the electron shell "dies out", and, consequently, in this case elastic scattering of a particle on an atom is entirely due to its interaction with the atomic nucleus:

$$\left(\frac{d\sigma}{d\Omega} \right)_{\text{el}} \Big|_{q \gg 1/a} \approx Z^2 \left(\frac{d\sigma}{d\Omega} \right)_R , \quad (2.51)$$

and, unlike (2.45) and (2.47), the cross section (2.51) is proportional to Z^2, instead of Z.

Now we substitute the obtained results into (2.40), having performed a change of the integration variable: $\vartheta \to q \approx k\vartheta$:

$$\langle \Theta^2 \rangle = n_0 x \int \frac{q^2}{k^2} \left[\left(\frac{d\sigma}{d\Omega} \right)_{\mathrm{el}} + \left(\frac{d\sigma}{d\Omega} \right)_{\mathrm{inel}} \right] \frac{2\pi q\, dq}{k^2}. \qquad (2.52)$$

It can now be clearly seen that the decisive contribution to the integral is given by elastic scattering involving large momentum transfers. We therefore drop the second term in (2.52) and substitute the approximate expression (2.51) into the first term. As a result we obtain

$$\langle \Theta^2 \rangle = n_0 x \frac{8\pi m_{\mathrm{a}}^2 Z_{\mathrm{a}}^2 Z^2 e^4}{\hbar^4 k^4} \int\limits_{1/a}^{q_{\max}} \frac{q^3 dq}{q^4} = n_0 x \frac{8\pi m_{\mathrm{a}}^2 Z_{\mathrm{a}}^2 Z^2 e^4}{(\hbar k)^4} \ln q_{\max} a. \qquad (2.53)$$

The quantity $q_{\max}$ has already been estimated above [see (2.32)].

By definition, the average angle of multiple scattering is calculated in accordance with

$$\Theta = \sqrt{\langle \Theta^2 \rangle}. \qquad (2.54)$$

We may single out the main dependence of Θ on the particle energy E, on its charge Z_{a}, and, also, on the parameters of the medium:

$$\Theta \sim \frac{Z_{\mathrm{a}} Z}{E} \sqrt{n_0 x}. \qquad (2.55)$$

2.4 Straggling: Fluctuations of Ionization Losses

The statistical character of the collisions of a particle with the atoms of a medium is not only manifest in the angular spread of the passing beam, but also in the spread of the total path ranges of individual particles and in the spread of the ionization losses occurring in a given layer of the medium. These phenomena are collectively called *straggling*.

From Fig. 2.4 it can be seen that the path range dealt with in Sect. 2.2 is actually the average range of the particle beam. The distribution of path

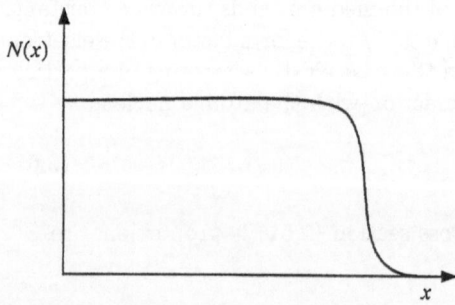

Fig. 2.4. Spread of path lengths of particles of given initial energy; $N(x)$ is the beam intensity upon transmission through a layer x

ranges of individual particles about the mean value $\langle R \rangle$ follows the Gaussian law quite well:

$$P(R)\mathrm{d}R = \frac{1}{\sqrt{2\pi\langle(\Delta R)^2\rangle}} \exp\left(-\frac{(R-\langle R\rangle)^2}{2\langle(\Delta R)^2\rangle}\right)\mathrm{d}R, \tag{2.56}$$

where $\langle(\Delta R)^2\rangle = \langle(R-\langle R\rangle)^2\rangle$ is the dispersion of this distribution. The relative straggling $\sqrt{\langle(\Delta R)^2\rangle}/\langle R\rangle$ for various fast heavy particles in various media is quite small: from some tenths up to one or two percent (Table 2.1). Its dependence on the initial energy of the particle correlates with the energy dependence of the stopping power of the medium: namely, up to the minimum of the $-\mathrm{d}E/\mathrm{d}x$ curve the quantity $\sqrt{\langle(\Delta R)^2\rangle}/\langle R\rangle$ falls slowly, as the particle energy rises, and further it increases slowly (Fig. 2.5). The relative spread of path ranges at a sole energy is smaller for α particles than for protons; this is the effect of the particle mass.

Table 2.1. Straggling $\sqrt{\langle(\Delta R)^2\rangle}/\langle R\rangle$
for 20 MeV protons in various media in %

Be	Al	Cu	Pb
1.3	1.5	1.6	1.8

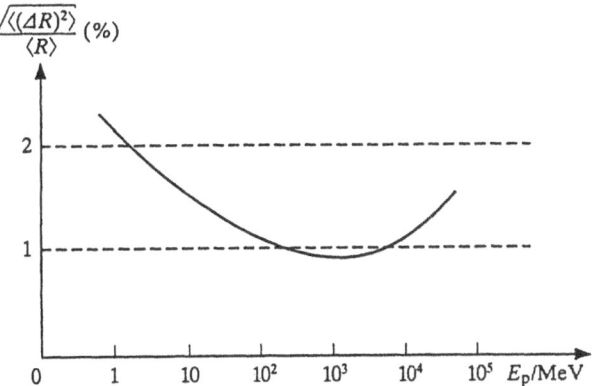

Fig. 2.5. Straggling for protons in aluminium versus the initial energy of the particles

Now, consider fluctuations of ionization losses in a given layer of a medium. If a chosen layer Δx is sufficiently thick for the particle to experience a large number of inelastic collisions, then the spread of its energy losses about their average value,

$$\langle\Delta E\rangle = \left(-\frac{\mathrm{d}E}{\mathrm{d}x}\right)\Delta x, \tag{2.57}$$

turns out to follow a Gaussian distribution:

$$P(\Delta E) = \frac{1}{\sqrt{2\pi}\sigma} \exp\left(-\frac{(\Delta E - \langle \Delta E \rangle)^2}{2\sigma^2}\right). \tag{2.58}$$

The dispersion of this distribution, σ^2, can be estimated as follows. The maximum energy lost in one collision by a particle of velocity v_a is determined by the maximum momentum transfer q_{max} (2.32):

$$\varepsilon_{max} = \frac{(\hbar q_{max})^2}{2m_e} = 2m_e v_a^2. \tag{2.59}$$

This quantity determines the order of magnitude of the mean particle energy lost in a single collision: $\langle \varepsilon \rangle \sim m_e v_a^2$. We shall divide the average energy (2.57) by it and obtain the order of magnitude of the mean number of inelastic collisions in a layer Δx:

$$\overline{N} \sim \frac{\langle \Delta E \rangle}{\langle \varepsilon \rangle} \approx \frac{4\pi Z_a^2 e^4}{(m_e v_a^2)} Z n_0 \Delta x \tag{2.60}$$

(we have dropped the logarithmic factor present in $-\mathrm{d}E/\mathrm{d}x$, since we are not interested in a quantitative precision of the estimates). The square root of (2.60) yields the root-mean-square deviation of the number of inelastic collisions from the mean value:

$$\Delta \overline{N} = \sqrt{\overline{N}} \sim \frac{2\sqrt{\pi} Z_a e^2}{m_e v_a^2} \sqrt{Z n_0 \Delta x}. \tag{2.61}$$

Hence we obtain the approximate value of the root-mean-square deviation from the mean value for the energy lost by the particle in a layer Δx:

$$\sigma \sim \Delta \overline{N} \langle \varepsilon \rangle \approx 2\sqrt{\pi} Z_a e^2 \sqrt{Z n_0 \Delta x}. \tag{2.62}$$

We have considered the case of a thick layer. If, on the contrary, the layer Δx is so thin that the average number of inelastic collisions experienced by the particle is small, then the spread of ionization losses no longer satisfy a Gaussian law (2.58). The distribution $P(\Delta E)$ in this case is no longer symmetric relative to $\langle \Delta E \rangle$, it is close to the Poisson distribution.

Third Lecture

3.1 "Dielectric Theory" of Ionization Stopping

The assumption that in passing through a medium a particle undergoes independent pair collisions with individual atoms turns out to be too inaccurate for examining the passage of particles through dense media. One of the effects not taken into account in this approximation is the polarization effect of the atoms of the medium under the influence of the electric field of the passing charged particle. Polarization of the medium weakens the influence of the electric field of the particle on the atomic electrons, so the energy transferred to them from the particle and, consequently, the stopping power of the medium is reduced. Hence it is seen that the stopping mechanism of charged particles in various media is somehow related to the dielectric properties of the medium. It is no chance that in studies of ionization stopping a significant place is occupied by the so-called *"dielectric theory"* of this phenomenon. Let us consider its main points.

Consider a heavy charged particle a travelling at a velocity $v(v \ll c)$ in a medium with a dielectric constant ε; the dispersion law $\varepsilon = \varepsilon(\omega)$, where ω is the angular frequency of the electric field, will be considered known. Thus, from optics and atomic physics we know that in a rarefied medium $\varepsilon(\omega)$ obeys the dispersion formula

$$\varepsilon(\omega) = 1 + \frac{4\pi e^2 n_0}{m_e} \sum_\nu \frac{f_\nu}{\omega_\nu^2 - \omega^2 - i\omega\gamma_\nu}, \tag{3.1}$$

where n_0 is the number of atoms per unit volume of the medium, ω_ν are the natural frequencies of the atoms of the medium, f_ν are the "weights" (oscillator strengths) of the respective transitions, and γ_ν are the damping coefficients of natural oscillations. The potential created by a moving particle 'a' is found from the general equations of electrodynamics:

$$\Delta\varphi(\boldsymbol{r}, t) = -4\pi\rho(\boldsymbol{r}, t), \tag{3.2}$$

where the charge density, according to our condition, has the form

$$\rho(\boldsymbol{r}, t) = Z_0 e \delta(\boldsymbol{r} - \boldsymbol{v}t). \tag{3.3}$$

It is easy to check by substitution that the following is the solution of (3.2):

$$\varphi(r,t) = \frac{1}{(2\pi)^3} \int \frac{4\pi Z_a e}{k^2} e^{ik\cdot(r-vt)} d^3k \,. \tag{3.4}$$

Hence it is seen that the electric field caused by the particle can be represented as a continuous set of monochromatic waves with frequencies calculated by the formula

$$\omega = k \cdot v \,. \tag{3.5}$$

If our particle 'a' were moving in empty space, the electric field strength created by it could be found from

$$E_0(r,t) = \nabla\varphi(r,t) = -\frac{1}{(2\pi)^3} \int \frac{4\pi Z_a e}{k^2} ik e^{ik\cdot(r-vt)} d^3k \,. \tag{3.6}$$

When the particle travels in a dielectric medium, this formula is changed:

$$E_0(r,t) \to E(r,t) = \frac{1}{(2\pi)^3} \int \frac{1}{\varepsilon(\omega)} \frac{4\pi Z_a e}{k^2} ik e^{ik\cdot(r-vt)} d^3k \,. \tag{3.7}$$

The stopping force exerted by the medium on the particle is the difference effect:

$$\begin{aligned} F(r = vt) &= Z_a e \big[E(r,t) - E_0(r,t) \big]_{r=vt} \\ &= -i\frac{1}{(2\pi)^3} \int \frac{4\pi Z_a^2 e^2}{k^2} k \left(\frac{1}{\varepsilon(\omega)} - 1 \right) d^3k \,. \end{aligned} \tag{3.8}$$

To simplify this integral we write down the differential as $d^3k = k^2 dk \sin\vartheta\, d\vartheta\, d\varphi$ (where ϑ is the angle between vectors k and v, and φ is the corresponding azimuthal angle), and using (3.5) we express ω through ϑ as $\omega = kv\cos\vartheta$. Then, integration over ϑ within the limits of $0°$–$180°$ may be replaced by integration over ω from $\omega = kv$ to $\omega = -kv$; here, $d\omega = -kv\sin\vartheta\, d\vartheta$. As a result, we find that the force $F(r = vt)$ has only a longitudinal component along the direction of motion of the particle, v, and its value is given by the expression

$$F(r = vt) = -i\frac{Z_a^2 e^2}{\pi v^2} \int\limits_0^\infty \frac{dk}{k} \int\limits_{-kv}^{kv} \left(\frac{1}{\varepsilon(\omega)} - 1 \right) \omega\, d\omega \,. \tag{3.9}$$

For computing the integral (3.9) it is necessary to know the dielectric constant of the medium, $\varepsilon(\omega)$, as a function of the frequency not only for positive but also for negative values of ω, i.e., including the nonphysical range of ω values. To this end one may take advantage of the general relationships known from optics for the real and imaginary parts of the dielectric constant $\varepsilon(\omega)$ as an analytic function of ω:

$$\begin{aligned} \operatorname{Re}\{\varepsilon(\omega)\} &= \operatorname{Re}\{\varepsilon(-\omega)\} \,, \\ \operatorname{Im}\{\varepsilon(\omega)\} &= -\operatorname{Im}\{\varepsilon(-\omega)\} \,. \end{aligned} \tag{3.10}$$

With these relations taken into account it can be seen that the stopping force (3.9) or, which is the same, the stopping power of the medium,

$-\mathrm{d}E/\mathrm{d}x = |F(r = vt)|$, is fully determined by the imaginary part of the dielectric constant:

$$-\frac{\mathrm{d}E}{\mathrm{d}x} = \frac{Z_{\mathrm{a}}^2 e^2}{\pi v^2} \mathrm{Im}\left\{ \int\limits_0^\infty \frac{\mathrm{d}k}{k} \int\limits_{-kv}^{kv} \frac{\omega\,\mathrm{d}\omega}{\varepsilon(\omega)} \right\}. \tag{3.11}$$

3.2 Application of the "Dielectric Theory": Ionization Stopping in a Rarefied Gas

Let us calculate the stopping power of a rarefied gas. From (3.1) it is seen that, in the case of a sufficiently low concentration of atoms, n_0, the dielectric constant $\varepsilon(\omega)$ differs little from unity. This means that the approximate expression

$$\frac{1}{\varepsilon(\omega)} \approx 1 - \frac{4\pi e^2 n_0}{m_{\mathrm{e}}} \sum_\nu \frac{f_\nu}{\omega_\nu^2 - \omega^2 - \mathrm{i}\omega\gamma_\nu}. \tag{3.12}$$

can be substituted for $1/\varepsilon(\omega)$ in (3.11). As a result, we obtain

$$-\frac{\mathrm{d}E}{\mathrm{d}x} \approx \frac{4\pi Z_{\mathrm{a}}^2 e^4 n_0}{m_{\mathrm{e}} v^2} \sum_\nu f_\nu \int\limits_{\omega_\nu/v} \frac{\mathrm{d}k}{k}. \tag{3.13}$$

We here took advantage of the relation

$$\int\limits_{-kv}^{kv} \frac{\omega\gamma}{(\omega_\nu^2 - \omega^2)^2 + \omega^2\gamma^2} \omega\,\mathrm{d}\omega \bigg|_{\gamma\to 0} = 0, \quad \text{if} \quad kv < \omega_\nu, \tag{3.14}$$

with the aid of which the lower limit of the integral over the variable k is determined. The physical meaning of this variable is the transfer of momentum to the atom of the medium resulting from the influence of the electric field of the moving charge. Formally, the integral $\int \mathrm{d}k/k$ in (3.13) is taken from $k_{\min} = \omega_0/v$ up to infinity. Actually (see Sect. 2.1), the momentum transferred to the atom in an inelastic collision is limited: $\hbar k_{\max} \approx 2m_{\mathrm{e}}v$. Substitution of this value into (3.13) yields

$$-\frac{\mathrm{d}E}{\mathrm{d}x} = \frac{4\pi Z_{\mathrm{a}}^2 e^4 n_0}{m_{\mathrm{e}} v^2} \sum_\nu f_\nu \int\limits_{\omega_\nu/v}^{2m_{\mathrm{e}}v/\hbar} \frac{\mathrm{d}k}{k}$$

$$= \frac{4\pi Z_{\mathrm{a}}^2 e^4 n_0}{m_{\mathrm{e}} v^2} \sum_\nu f_\nu \ln \frac{2m_{\mathrm{e}}v^2}{\hbar\omega_\nu}. \tag{3.15}$$

Now, we shall apply the *sum rule*, known from atomic physics, for the oscillator strengths:

$$\sum_\nu f_\nu = Z, \tag{3.16}$$

where, Z is the number of electrons in the atom. We define the mean excitation energy of the atom, I, by the relation

$$\ln I \equiv \frac{1}{Z} \sum_{\nu} f_{\nu} \ln(\hbar\omega_{\nu}),$$ (3.17)

and use it to express the stopping power of the medium:

$$-\frac{dE}{dx} = \frac{4\pi Z_a^2 e^4}{m_e v^2} Z n_0 \ln \frac{2m_e v^2}{I}.$$ (3.18)

Formally, (3.18) is identical to (2.34), derived above within the framework of the quantum theory of ionization stopping. The mean excitation energy of the atom, I, occurring in (2.34) remained undetermined; on the contrary, here it is given by relation (3.17) and, consequently, can be calculated in a straightforward manner if the distribution of the oscillator strengths in an atom of the medium is known. In Sect. 2.1 it has already been noted that calculations performed for various atoms always reveal the mean energy I to be higher than the ionization potential of the atom.

3.3 Stopping of a Charged Particle in a Degenerate Electron Gas

The problem, we are about to consider is widely applied in theoretical studies of the passage of charged particles through plasma and condensed media. In this aspect it is quite the opposite of the problem we have just dealt with in Sect. 3.2: therein we diverted our attention from any collective effects relevant to the passage of particles through matter, whereas now the collective stopping mechanism will occupy a most important place.

From atomic physics it is known that the main characteristics of a degenerate electron gas are the Fermi boundary (defined as the limit energy of an electron ε_F or its momentum k_F) and the frequency of collective ("plasma") oscillations, ω_0 – the "plasma frequency". Both these characteristics are expressed in terms of the electron density n_e:

$$k_F = (3\pi^2 n_e)^{1/3};$$ (3.19)

$$\omega_0 = \sqrt{\frac{4\pi e^2 n_e}{m_e}}.$$ (3.20)

For orientation, we note that the average electron density n_e in metals is of the order of $10^{23}\,\mathrm{cm}^{-3}$, i.e., of the order of unity in atomic units. From (3.19) and (3.20) it follows that, in this case, k_F and ω_0 are also of the order of several atomic units, i.e., k_F and ω_0 are of the order of $10\,\mathrm{eV}$.

According to "dielectric theory", the stopping power of any substance is determined by its dielectric constant. We shall not derive the dependence

$\varepsilon(\omega)$ on parameters characterizing the state of an electron gas and the charged particle passing through it. We shall, however, note the main points in obtaining the final formula. In Sect. 3.1 it was shown that a charged particle carries with it a "packet" of photons exhibiting special properties in relation to the quanta of the free electromagnetic field: the frequency ω and wave vector $\boldsymbol{k}$ are not related to each other by the usual relation $\omega = kc$. These are so-called *virtual photons*. For them, however, like for ordinary photons, the law of momentum conservation is obeyed in interactions with electrons of the medium:

$$\boldsymbol{k}_e' = \boldsymbol{k}_e + \boldsymbol{k}, \tag{3.21}$$

where $\boldsymbol{k}_e$ and $\boldsymbol{k}_e'$ are the initial and final momenta of a free electron of the medium. Hence the energy transferred from a particle to the medium, when a virtual photon with a wave vector $\boldsymbol{k}$ is absorbed by an electron with an initial momentum $\boldsymbol{k}_e$, is

$$\Delta E = \frac{\hbar^2 k_e'^2}{2m_e} - \frac{\hbar^2 k_e^2}{2m_e} = \frac{\hbar^2}{2m_e}\left[(\boldsymbol{k}_e + \boldsymbol{k})^2 - k_e^2\right] \tag{3.22}$$

(by the way, note that absorption of an ordinary, real photon by a free electron is impossible, since the laws of momentum and energy conservation cannot be satisfied simultaneously in this case; see Sect. 7.1 for details).

Here, we must recall the Pauli principle: an electron in a degenerate electron gas can undergo transition from an initial state $|\boldsymbol{k}_e\rangle$ to another state $|\boldsymbol{k}_e'\rangle$ only if this final state lies beyond the Fermi sphere, i. e., if $k_e' > k_F$. We now turn to the general formula for the dielectric constant, (3.1). In applying it to the case of a degenerate electron gas, we substitute a continuous set of values $\Delta E/\hbar$, where ΔE is taken from (3.22), for the spectrum of eigenfrequencies ω_ν; the electron density is substituted for the number of atoms, n_0, per unit volume of the medium, and the corresponding probabilities of one-electron transitions, $f(\boldsymbol{k}_e \to \boldsymbol{k}_e')$ are substituted for the oscillator strengths f_ν. Here, summation over ν is replaced by integration over all the initial electron states, i. e., over all the states inside the Fermi sphere. Taking the conservation law (3.21) and the Pauli principle,

$$\varepsilon(k,\omega) = 1 + \frac{4\pi e^2 n_e}{m_e} \int \left. \frac{f(\boldsymbol{k}_e \to \boldsymbol{k}_e + \boldsymbol{k})\mathrm{d}^3 k_e}{\frac{\hbar^2}{4m_e^2}\left|(\boldsymbol{k}_e + \boldsymbol{k})^2 - k_e^2\right|^2 - \omega^2 - i\omega\gamma}\right|_{\substack{k_e \leq k_F \\ |\boldsymbol{k}_e + \boldsymbol{k}| > k_F}} \tag{3.23}$$

into account provides for all possible final states being taken into account also [note that the whole combination of factors before the integral is nothing but the square of the plasma frequency (3.20)]. It remains for us to substitute here the appropriate expression for the probability of an one-electron transition, $f(\boldsymbol{k}_e \to \boldsymbol{k}_e')$ (it is given by quantum electrodynamics), to perform integration over $\mathrm{d}^3 k_e$ taking the restrictions indicated into account, and then to substitute the expression obtained for the dielectric function $\varepsilon(k,\omega)$ into the integral in (3.11). Generally, this entire integration procedure can be fulfilled only numerically. As to the analytical formulae, they can be obtained only approximately and, even in such cases, only if certain relations exist between various parameters of the problem.

Thus, if the velocity of the passing particle is much greater that the velocity of an electron on the Fermi surface, i.e., $v \gg v_F = \hbar k_F/m_e$, the stopping power of the degenerate electron gas is calculated by

$$-\frac{dE}{dx} = \frac{4\pi Z_a^2 e^4}{m_e v^2} n_e \left[\ln \frac{2 m_e v^2}{\hbar \omega_0} - \frac{3}{5}\left(\frac{v_F}{v}\right)^2 - \frac{3}{14}\left(\frac{v_F}{v}\right)^4 + \dots \right], \qquad (3.24)$$

where higher-order terms are omitted.

We have on hand a close resemblance between the obtained expression and the already familiar formulae for the stopping power of a medium. Comparing it, for instance, with (2.34) we see that the energy of plasma oscillations, $\hbar\omega_0$, plays the same part in (3.24) as the mean excitation energy of an atom, I, does in formula (2.34). The simple physical meaning of the result presented is revealed: a fast charged particle passing through a degenerate electron gas is slowed down owing to the excitation of collective plasma oscillations in the gas.

One must not forget that (3.24) is valid only under the condition that $v \gg v_F$. If this condition is not fulfilled we write the stopping power of a degenerate electron gas in the form suggested by expression (3.24):

$$-\frac{dE}{dx} = \frac{4\pi Z_a^2 e^4}{m_e v^2} n_e L\left[n_e, v\right]. \qquad (3.25)$$

By explicitly writing out the arguments n_e and v in the last factor $L \equiv L[n_e, v]$ we stress that the electron gas density n_e enters into $-dE/dx$ not only as a trivial proportionality factor, but also, in a more complex way, via $L[n_e, v]$ and, for instance, via the plasma frequency $\omega_0 = \omega_0(n_e)$:

$$L\left[n_e, v\right] \approx \ln \frac{2 m_e v^2}{\hbar \omega_0}. \qquad (3.26)$$

Note that the dependence of the factor $L[n_e, v]$ upon the electron density weakens the rise of the stopping power of matter as n_e increases. Here is a direct relation to the so-called "density effect" to be dealt with in Sect. 4.1.

3.4 Local Density Approximation

In a medium with an inhomogeneous electron density distribution, the energy losses of a particle depend strongly on the trajectory of its motion, which is especially obvious in the case of the slowing down of particles channelled in crystals (see Sect. 5.1). To take into account the specific features of the passage of charged particles through inhomogeneous media one applies the so-called "local density approximation". It is formulated as follows.

Let the electron density distribution along the trajectory of the particle's motion be known: $n_e = n_e(r)$. We shall consider the energy-loss rate in the vicinity of each point r to depend only on the density at that point, i.e., on the "local density":

$$-\frac{\mathrm{d}E}{\mathrm{d}x} = \frac{4\pi Z_a^2 e^4}{m_e v^2} n_e(r) L[n_e(r), v] . \qquad (3.27)$$

Then the energy losses at any finite segment of the trajectory are given by the integral

$$(-\Delta E)_{\Delta s} = \int\limits_{\Delta s} \frac{4\pi Z_a^2 e^4}{m_e v^2} n_e(r) L[n_e(r), v] \, \mathrm{d}s . \qquad (3.28)$$

As we see, the local density approximation is essentially quite simple, and if one assumes the formulae of the theory of a degenerate electron gas to be valid for $L[n_e(r), v]$, then the whole procedure of calculating energy losses also turns out to be quite simple. Owing to this fact, the local density approximation is widely applied, even for computing the energy losses of a particle colliding with a single atom. This approach is adopted, for example, in studies of the dependence of such losses on the collision impact parameter (Fig. 3.1). It must, however, be borne in mind that from a theoretical standpoint the legitimacy of this approximation as applied to various real media has not always been sufficiently investigated.

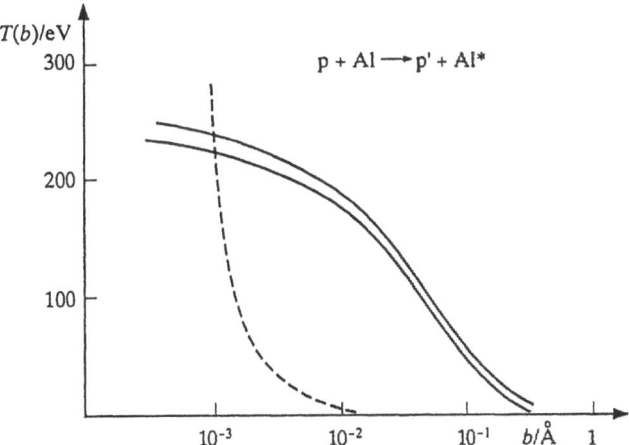

Fig. 3.1. The mean ionization losses $T(b)$ of a 7 MeV proton scattered by an aluminium atom – calculated from (3.28) for a straight-line trajectory of the proton [3.1]. The *two solid lines* are for two different choices of the electron density $n_e(r)$; for orientation, the *dotted curve* shows the energy loss of a proton undergoing Rutherford scattering from a nonscreened atomic nucleus

3.5 Relativistic Effects: The Bethe–Bloch Formula

According to the nonrelativistic theory of ionization stopping, the maximum momentum that can be transferred from a heavy particle of mass $m_a \gg m_e$ to an electron is $2m_e v$ (2.32). Relativistic kinematics gives another result for this quantity:

$$(p_e)_{max} = \frac{2m_e v}{\sqrt{1 - \beta^2}}, \tag{3.29}$$

i. e., the region of integration over the momentum transferred, when the effective stopping $S(E)$ is calculated, is extended in comparison with what is given by (2.27–2.33), and taking relativism into account leads to enhancement of the effective stopping as compared with predictions of nonrelativistic theory. The physical reason for such an enhancement is the effect of relativistic "compression" of the electric field of a passing particle: in the direction of flight of the particle the field is reduced by a factor of $(1 - \beta^2)$, whereas in the transverse direction it is enhanced by a factor of $1/\sqrt{1 - \beta^2}$, and thus influences a greater number of the electrons of the medium than would be possible if the field were spherically symmetric.

Taking relativism into account consistently yields the following result for the stopping power:

$$-\frac{dE}{dx} = \frac{4\pi Z_a^2 e^4}{m_e v^2} Z n_0 \left[\ln\left(\frac{2m_e v^2}{I}\right) + \ln\left(\frac{1}{1 - \beta^2}\right) - \beta^2 \right]. \tag{3.30}$$

At $v \ll c$ this formula transforms into (2.34), and in the relativistic limit of $v \to c$ it exhibits a constant, although slow, rise in the stopping power. Actually, such an infinite rise is not observed: it is hindered by the "density effect", the screening of the influence the field of the passing particle has on the distant atoms of the medium, which is due to the atoms nearest to the particle being polarized. To take the density effect into account, one more correction is introduced in formula (3.30):

$$-\frac{dE}{dx} = \frac{4\pi Z_a^2 e^4}{m_e v^2} Z n_0 \left[\ln\left(\frac{2m_e v^2}{I}\right) + \ln\left(\frac{1}{1 - \beta^2}\right) - \beta^2 - \delta \right]. \tag{3.31}$$

The term δ, the explicit dependence of which on the parameters is given by different authors differently, reflects the restrictive action of the polarization of the medium.

This effect is insignificant in gases, and the logarithmic rise in the energy dependence of the stopping power is observed up to extremely high energies. In solids and liquids, where the electron density is high, the polarization effect is more pronounced. In condensed media, the $-dE/dx$ curve reaches a minimum at approximately $E = m_a c^2$, and then undergoes an insignificant rise practically reaching a plateau (Fig. 3.2).

Formula (3.31), the main formula in the theory of ionization stopping of heavy fast charged particles, is called the *Bethe–Bloch formula*.

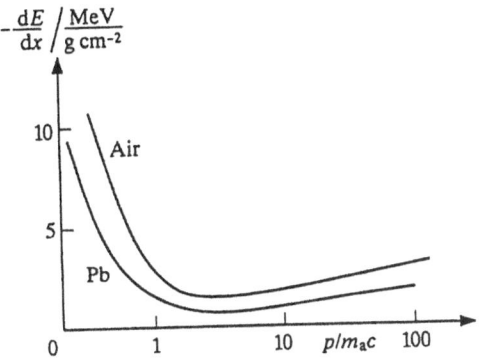

Fig. 3.2. Mass stopping power of air and lead for singly charged particles

Fourth Lecture

4.1 Ionization Stopping of Slow Particles

In this lecture we shall briefly consider several additional issues to complete the general picture of stopping mechanisms of charged particles in matter and of the methods for their theoretical description, presented in the three preceding lectures. We shall first deal with the ionization stopping of slow particles.

The Bethe–Bloch formula describes ionization losses of particles with velocities exceeding the mean velocity of electrons in atoms of the medium (1.27). From Fig. 2.2. it can be seen that the $-dE/dx$ curves have maxima at velocities of the order of $\langle v_e \rangle$. The stopping power of the medium decreasing toward lower velocities reflects the corresponding behavior of the excitation and ionization cross sections of atoms in ion–atom collisions (Fig. 4.1). Theoretical description of such processes at small ion velocities requires more complex methods than perturbation theory, of which we hitherto took advantage in dealing with fast particles. At small velocities it no longer suffices to take into account the influence of the field exerted on an atom by the incident particle in the lowest order of perturbation theory when the wave functions of an isolated atom are taken to be the base functions. Here, it is

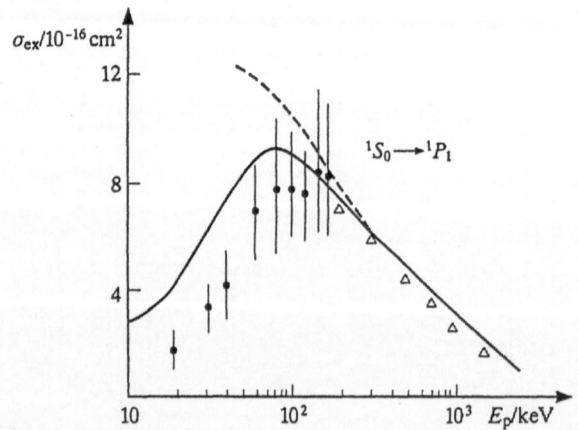

Fig. 4.1. Excitation cross section versus proton energy for the $2p^5 3s : {}^1P_1$ level in a neon atom when excited by an incident proton [4.1]. The *solid line* is calculated with account taken of perturbation of the electron shell of the atom during the collision; the *dotted line* is calculated in the Born approximation; the *symbols* are experimantal results

necessary to construct a totally different set of basis wave functions of the atomic electrons, which would consistently take into account the two-center character of the electric field of heavy particles (the atomic nucleus and the incident particle) and the change in its configuration occurring as these particles approach each other and then move apart. In quantum collision theory, such approaches have been elaborated upon and yield good results when modern numerical methods are applied. Our task does not involve examination of these theoretical methods; it is only important to clarify that at velocities lower than atomic velocities, the stopping mechanism, although essentially remaining an "ionization" mechanism (i. e., as in the case of large velocities the particle energy is spent on excitation and ionization of atoms), becomes quite different, and the dependence of the stopping on the particle velocity becomes opposite to the dependence observed for fast particles. If we trace, in our mind, the stopping process of a particle down to "the end", we will reach energies of the order of 1 eV when the particle is no longer capable of ionizing or exciting atoms of the medium. Further stopping of such particles is due only to elastic collisions. This process gradually transforms into diffusion, and, ultimately, the charged particles that happened to land in the medium start to participate in the common thermal motion of the atoms of the medium.

One of the first theoretical works on the passage through matter of charged particles with velocities lower than the atomic velocity was the work of Fermi and Teller (in 1947), in which the dependence of the stopping power on the particle velocity was established with the aid of the Fermi-gas model. Consider a heavy charged particle moving in a degenerate electron gas with a velocity v that is much smaller than the Fermi velocity of an electron in this gas: $v \ll v_F$. In colliding with electrons, our particle may transfer to them part of its energy, but only if such a transfer results in the final electron velocity exceeding v_F; all the states inside the Fermi sphere are occupied. Thus, owing to the Pauli principle, of all the electrons filling up the Fermi sphere, only a small fraction can take part in stopping the passing charged particle; this fraction is smaller, the smaller the velocity of the particle, v. For quantitative estimates we shall apply the result obtained by Fermi and Teller:

$$-\frac{dE}{dx} = \frac{2}{2\pi} \frac{m_e^2 Z_a^2 e^4 v}{\hbar^2} \ln \frac{v_F}{\alpha c},$$

(4.1)

where α is the fine-structure constant. As one can see, the energy losses of a particle are proportional to its velocity and depend very weakly on the electron-gas density.

We shall present one more formula derived within the framework of the Thomas–Fermi model for calculating ionization losses in the case of small particle velocities:

$$-\frac{dE}{dx} = 8\pi e^2 a_0 Z_a^{1/6} \frac{Z_a Z}{(Z_a^{2/3} + Z^{2/3})^{3/2}} n_0 \frac{v}{v_B}.$$

(4.2)

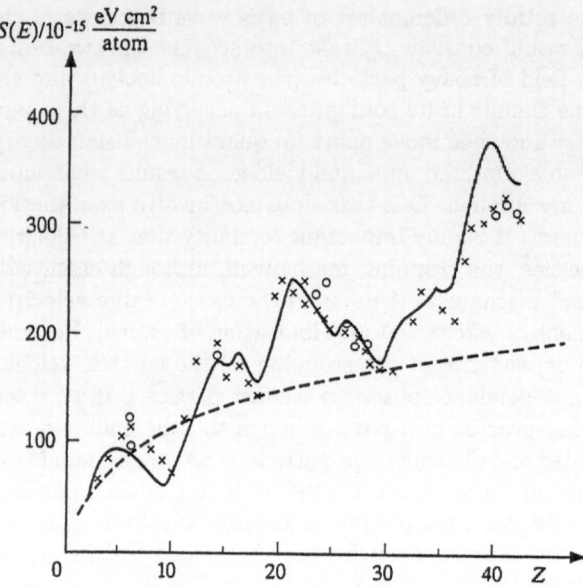

Fig. 4.2. Effective stopping of 800 keV N$^+$ ions versus the atomic number of the medium. The *solid line* is a phenomenological fit to experimental data obtained by different groups; the *dotted line* is calculated from the Thomas–Fermi model; the *symbols* are experimental results [4.2]

Here a_0 and $v_B = \alpha c$ are the Bohr radius of the atom and the Bohr velocity of the atomic electron, respectively. As in the case of descriptions of other atomic characteristics, the Thomas–Fermi model, being a statistical model, conveys only the general tendency of the dependence of the stopping power on the atomic number of the medium. Thus, for instance, (4.2) conceals the Z oscillations characteristic of the stopping power, which have been experimentally established and reflect the periodicity in properties of the chemical elements (Fig. 4.2).

The above formulae describe the ionization losses to the left of the maximum in a plot like Fig. 4.1 or Fig. 2.2 and the Bethe–Bloch formula describes the same curve to the right of the maximum. A consistent theoretical description of stopping in the region of the maximum itself turns out to be very complex.

4.2 Bragg's Composition Law

Let us again turn to the Bethe–Bloch formula describing ionization stopping of fast particles. How can it be generalized to the case of a medium containing atoms of different sorts?

For this purpose, Bragg, at the beginning of the century, put forward a simple *"composite law"* (it is also called the Bragg rule): if a substance represents a chemical compound $A_m B_n$ composed of atoms A and B, then its stopping power is the weighted sum of the stopping powers of the constituent chemical elements:

$$\left(-\frac{\mathrm{d}E}{\mathrm{d}x}\right)_{A_m B_n} = m\left(-\frac{\mathrm{d}E}{\mathrm{d}x}\right)_A + n\left(-\frac{\mathrm{d}E}{\mathrm{d}x}\right)_B . \tag{4.3}$$

The Bragg rule follows from the assumption that the chemical bound between atoms in the molecules does not affect the interaction of the particle with the atomic electrons as it passes through the medium. Actually, this is not so, since the atoms of a substance always exhibit altered individual properties when they form molecules. It may be assumed that the extent to which the Bragg rule is violated depends on the relationship between the corresponding contributions to the stopping power from the electrons of the outer atomic shells (participating in the chemical bound) and the electrons belonging to the inner shells (feeling this bound weakly). In the preceding sections, including those presenting derivations of formulae such as the Bethe–Bloch formula, we did not need to pay attention to this aspect of the issue, since in the conditions of a disordered monoatomic medium, when the stopping power of the medium depends on the effective stopping power of each individual atom, everything reduces to its mean excitation energy I, within which the contributions of separate shells level out. However, in modern studies the issue of the relationship between the contributions of inner and outer shells to the effective stopping power gives rise to strong interest. This interest is associated not only with investigations of the stopping of particles in molecular media, but in particular with the issue of ionization losses of particles undergoing channelling (see Lecture 5), the mechanism of which is highly selective with respect to the contribution of individual atomic shells to the stopping.

Experiments show that in the case of binary compounds such as $A_m B_n$ the stopping power is always several percent smaller than that expected from the composite law (4.3).

4.3 The Stopping Power of a Medium for Particles and Antiparticles: The Z^3 Correction to the Bethe–Bloch Formula

In modern physics, at the boundary between elementary particle physics, nuclear physics, and atomic physics, two large complex programs have arisen: the program of muon catalysis of nuclear fusion (see Lecture 14) and the program of studies with low-energy antiprotons, within the framework of which many issues of the interaction of charged particles with matter have been newly raised. One such issue concerns the relationship between the stopping powers of matter for particles and antiparticles.

The Bethe–Bloch formula, like the entire stopping theory of fast particles, with which we have become familiar above, yields no effects depending on the sign of the particle charge Z_a. Within the framework of classical theory (see Sect. 1.2) this independence from the sign of the charge is determined by the choice of the straight-line trajectory of motion of the particle, and by the fact that we have neglected the displacement of electrons in the medium due to their interaction with the particle. In these conditions, the energy transferred to the electron, $T_e = 2Z_a^2 e^4/(b^2 m_e v_a^2)$ [see (1.31)], is the same independently of whether the electron interacts, for instance, with an antiproton or with a proton. Within quantum theory (see Sect. 2.1) the independence of the stopping power from the sign of the charge is due to perturbation theory – the plane-wave Born approximation – underlying its examination. Thus, the absence of any effect due to the sign of the charge in our description of the passage of a charged particle through matter results from certain simplifying assumptions made earlier.

Outside the framework of these assumptions a number of reasons can be immediately found for why inelastic collisions of particles and antiparticles with the atoms of a medium start to differ. Here are some of these reasons: (a) Coulomb bending of the trajectory of motion of a particle (away from the atomic nucleus for a positively charged particle, and approaching the nucleus in the case of a negatively charged particle); (b) effective enhancement of the binding of atomic electrons when the nucleus is approached by a proton or by any other positively charged particle (the charge at the center of the atom, which "feels" the atomic electrons, seems to increase by unity), and a similar drop in this coupling when a negatively charged particle approaches the nucleus; (c) the presence of the competing charge-exchange process (capture of the atomic electron) when the atom is excited by a positively charged particle, and the absence of such a process in when a negatively charged particle is involved. Note that all three reasons act in the same direction, i. e., "in favor" of negatively charged particles: the excitation and ionization cross sections of atoms by antiprotons $\overline{p}$, as well as by μ^- and π^- mesons, should exceed the respective cross sections for protons p, μ and π mesons.

What is the actual situation? It is not quite clear yet, since high-level experiments capable of providing an answer to this question have only recently become feasible. Considerable interest was shown in data on the interaction of antiprotons with atoms, obtained five years ago at the CERN low-energy antiproton storage ring (LEAR). Some of these results are presented in Fig. 4.3 together with earlier data for protons. Especially surprising are the data relevant to the cross sections of double ionization: these cross sections are much higher in the case of antiprotons than for protons.

No similar data on the ionization cross sections for muons $\mu^\pm$ and pions $\pi^\pm$ are available. Although quite incomplete, experimental data on parameters characterizing the actual passage of these particles through matter do exist. Thus, back in the 1960s, it was established that at moderate particle velocities ($\langle v_\pi \rangle/c \approx 0.5$–$0.7$) the stopping power of matter for π^+ mesons is somewhat higher than for π^- mesons, whereas the total path ranges of π^-

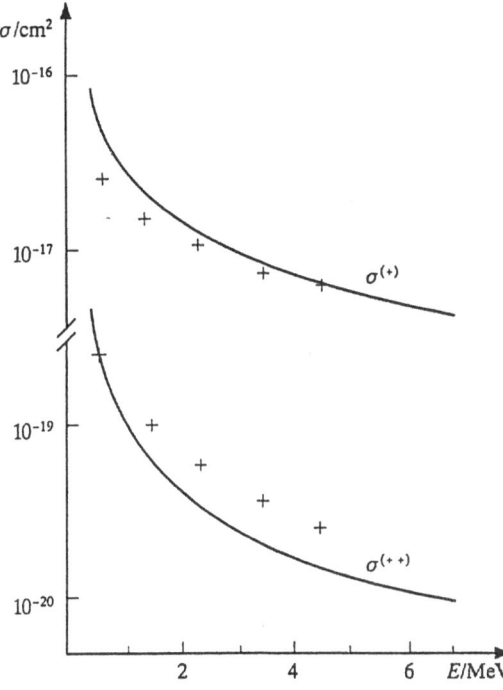

Fig. 4.3. Single $(\sigma^{(+)})$ and double $(\sigma^{(++)})$ ionization cross sections for helium atoms ionized by protons (*solid curves*) and antiprotons (*crosses*) [4.3]

mesons are somewhat larger than those for π^+ mesons. We shall see that this is the opposite to what is observed in the case of ionization cross sections due to protons and antiprotons and apparently disproves the qualitative explanations of the charge-sign effect presented above. Truly, other arguments can be presented from which it follows that the effect of the charge sign should manifest itself in favor of the positively charged particles. Indeed, a positively charged particle passing through the "gas" of electrons in a medium attracts them, so that along the path of the particle there will always be created a region of enhanced electron density, and this, in accordance with conventional ionization stopping theory, favors a rise in the energy losses of the particle. In the case of a negatively charged particle its polarizing influence on the electrons of the medium is of the opposite sign: the particle apparently gives rise to a certain rarefication of the electron gas along its trajectory, which results in its stopping being less strong.

It was noted above that it is still not quite clear how the sign of the particle's charge affects the parameters characterizing its passage through matter. One might think that in a transition from high to low velocities the effect itself changes sign, i.e., the systematic excess of inelastic interaction cross sections for positively charged particles over that for cross sections for negatively charged particles, expected at high velocities, changes when transition to low velocities takes place, in such a manner that the excitation and ionization cross sections due to positively charged particles become smaller than the corresponding cross sections for negatively charged particles.

The question at issue is related to the problem of the so-called "Z^3 correction" to the Bethe–Bloch formula. What is meant is the general formula (3.31), which we shall now give in a more compact form:

$$-\frac{dE}{dx} = \frac{4\pi Z_a^2 e^4}{m_e v^2} Z n_0 L. \tag{4.4}$$

Bear in mind that the calculation of the factor L (the "ionization logarithm") is performed somewhat differently in different approaches. It is essential, however, that in all the approaches in Lecture 3 the stopping power $(-dE/dx)$ is always proportional to the square charge of the particle, Z_a^2. In the general case, however, L is written in the form

$$L = L_0 + Z_a L_1 + Z_a^2 L_2 + \dots, \tag{4.5}$$

where L_0 corresponds to our previous formulae for L, whereas L_1 provides the correction to $-dE/dx$, which is proportional to the cubic charge of the particle (the "Z^3 correction"). Calculations performed by various authors applying different methods reveal the correction L_1 to be positive. This sign of the Z^3 correction is in agreement with the effect due to the charge sign, considered above for the passage through matter of fast protons, π^+ pions and μ^+ muons ($Z_a = +1$), and antiprotons, π^- pions and μ^- muons ($Z_a = -1$).

To test the ideas presented above relative to the Z^3 correction one can take advantage of available experimental data on the stopping powers of media for protons and α particles. In accordance with the above exposition, the stopping power for fast α particles, $(-dE/dx)_\alpha$, should exceed the stopping power exhibited by the same medium for protons, $(-dE/dx)_p$ by somewhat more than just a factor of 4, as expected from the Bethe–Bloch formula:

$$\left(-\frac{dE}{dx}\right)_\alpha \Big/ \left(-\frac{dE}{dx}\right)_p \bigg|_{v_\alpha = v_p} > \frac{Z_\alpha^2}{Z_p^2} = 4. \tag{4.6}$$

This is precisely what it turns out to be. In Fig. 4.4, experimental data are presented for tantalum and aluminium. We draw attention to the fact that the Z^3 correction decreases monotonously as the velocity of the particle rises. In this connection we note that as the energy of the incident particle increases, the number of atomic electrons for which the binding to the nucleus becomes less and less significant in the course of collision with the passing particle rises: the particle interacts with them as if they were free electrons. And in this case the Rutherford formula "works" more and more precisely, indicating a rigorous Z_a^2 dependence for the collision probability, without any corrections. Taking into account that the mean binding energy of electrons in an atom increases with the atomic number of the medium, we find that the above arguments permit us to understand why, as shown in Fig. 4.4, the Z^3 correction is larger for tantalum, which is the heavier of the two elements, than for aluminium.

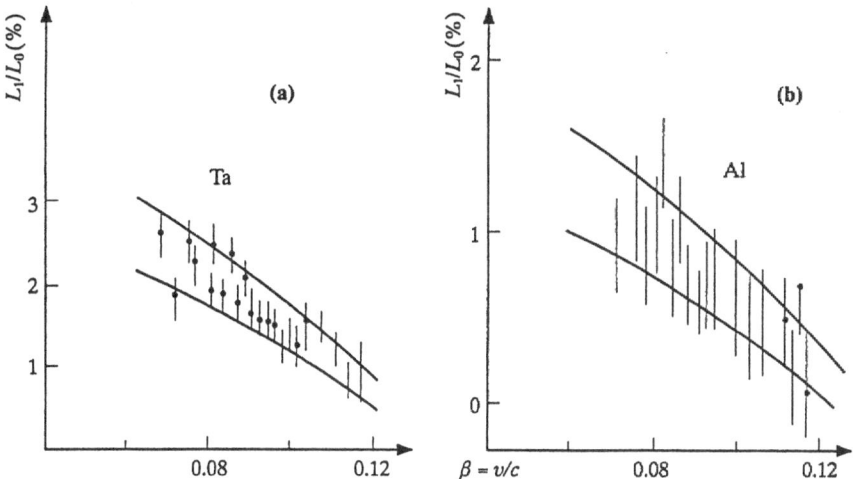

Fig. 4.4a, b. Parameter of Z^3 correction, L_1/L_0, for tantalum (a) and aluminium (b) from data on the stopping powers of these substances for α particles and protons. The *solid curves* show the corridor of predictions based on various versions of the theory of ionization stopping [4.4]

Turning to (4.5) one may think that the Z^3 correction is more readily revealed not by comparison between the stopping powers of matter for protons and α particles, but by that for protons and some heavier nuclei carrying a larger charge Z_a. Actually, this turns out to be quite difficult to do: the character of the passage through matter of multicharge particles differs essentially from what we know for particles with small charges, so that straightforward application of formulae such as the Bethe–Bloch formula in such cases turns out to be impossible, even with corrections (4.5).

4.4 Passage of Multiple-Charged Ions Through Matter

Consider a fast particle 'a' of charge $Z_a \gg 1$ passing through a medium. The familiar processes of excitation and ionization of the atoms of a medium, taking place when such a particle interacts with the atomic electrons, are accompanied by intense capture processes of these electrons. If $Z_a \gg 1$, such states of the a + e system will always be found, transition to which (at least of the outer electrons of the atoms of the medium) is favored from the point of view of energy (Fig. 4.5). The probability of resonance processes ($\varepsilon_f \approx \varepsilon_i$) is also significant when capture proceeds not to the ground or any other low-lying state but to highly excited states of the a + e system. The process described results in alteration of the charge state of the particle passing by, and it is conventionally called *charge exchange*.

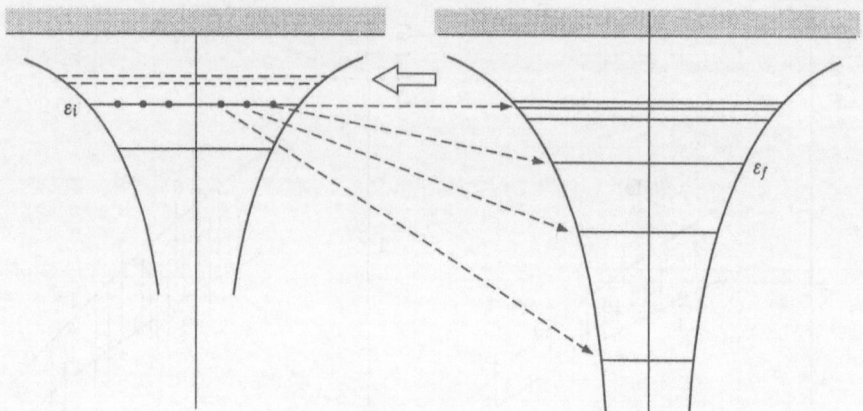

Fig. 4.5. Scheme of charge exchange occurring when a multicharged ion passes through a substance: on the *left* are the energy levels of the external electrons in the atom of the substance, on the *right* are the levels in the ion–electron system

   Imagine the particle starting its motion in the medium as a "bare" ion of charge Z_a. Owing to charge exchange, a cloud of bound electrons gradually grows up around it. As the number of such electrons increases, they occupy more and more orbits in the field of the charge Z_a, which results in the inverse process – the passing ion losing a part of the acquired electrons (so-called *stripping*) – becoming more significant. The combination of capture and stripping makes the charge exchange process exhibit a dynamic character in the case of multicharge ions, and when a beam of such ions passes through a medium we always have to deal with a whole set of various charge states of the ion.

   Let the quantity $N(Z'_a, x)$ characterize the intensity of the fraction of charge Z'_a in this composite beam at a distance x from the entrance of the beam to the medium. The distribution of various charge states in the beam may be characterized by weight coefficients

$$p(Z'_a, x) = \frac{N(Z'_a, x)}{\sum_{Z'_a} N(Z'_a, x)}, \qquad (4.7)$$

and via these coefficients it is possible to determine the mean charge and the dispersion of the charge distribution:

$$\overline{Z}_a(x) = \sum_{Z'_a} Z'_a p(Z'_a, x),$$

$$D_{Z_a}(x) = \sum_{Z'_a} \left[Z'_a - \overline{Z_a(x)} \right]^2 p(Z'_a, x), \qquad (4.8)$$

all for any arbitrarily chosen x.

If $\overline{\sigma}_{ex}$ is the mean charge exchange cross section, then the quantity

$$l_{ex} = (\overline{\sigma}_{ex} n_0)^{-1}, \tag{4.9}$$

where n_0 is the density of the atoms in the medium, is the mean path the beam must cover before the distribution of charge states in the beam is stabilized, i.e., a dynamic equilibrium is established between the capture and loss of electrons by an ion. The equilibrium mean ion charge,

$$\overline{Z}_a(\infty) = \sum_{Z_a'} Z_a' p(Z_a', x \gg l_{ex}), \tag{4.10}$$

and the dispersion of the equilibrium distribution are determined accordingly.

The equilibrium distribution of charge states of multicharge ions is seen to be established from experiments in which the passage of multicharge ions through matter is studied, given various initial ion charges Z_a^{in} – both above and below the equilibrium value – at the entrance to the target. When $Z_a^{in} \gg \overline{Z}_a(\infty)$, the mean charge of the ions, $\overline{Z}_a(x)$, falls monotonously as they pass through the medium, and when $Z_a^{in} \ll \overline{Z}_a(\infty)$ it rises monotonously and tends toward one and the same limit (Fig. 4.6). Note that the higher the

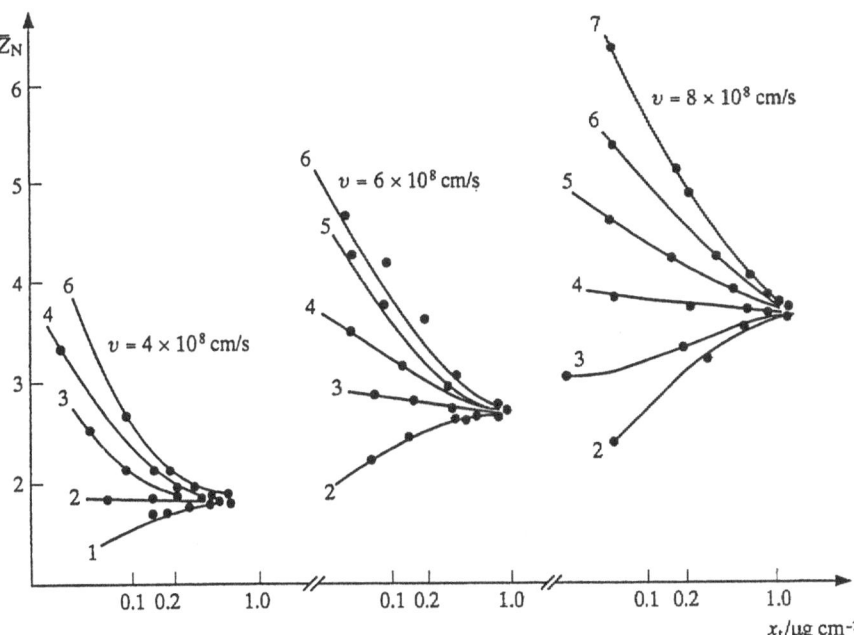

Fig. 4.6. Mean charge of nitrogen ions passing through a nitrogen gas target N_2 [4.5]. For clarity, curves are drawn through experimental points, the numbers near the curves indicate the initial charge Z_N^{in}; the horizontal axis represents the target thickness x_t

velocity of an ion, the easier it loses its electron instead of capturing an electron from an atom of the medium, i. e., the more the dynamic equilibrium shifts toward large values of the equilibrium charge $\overline{Z}_a(\infty)$.

The charge exchange process strongly affects the stopping of multicharge ions in matter. Thus, if a beam of "bare" ions of charge Z_a impinges upon the target, then the decrease in its mean charge $\overline{Z}_a(x)$, occurring as the ions penetrate the medium, leads to a weakening in the interaction of the ions with the atoms of the medium. Although the stopping power of a medium for a "bare" multicharge ion should be greater than that for a proton of the same velocity, by a factor of Z_a^2, the stopping power for a multicharged ion is actually significantly smaller:

$$\left(-\frac{\mathrm{d}E}{\mathrm{d}x}\right)_a < Z_a^2\left(-\frac{\mathrm{d}E}{\mathrm{d}x}\right)_p. \tag{4.11}$$

In the characterization of the stopping of multicharge ions the special concept of the *effective charge* of an ion is introduced, which is close, but not identical to the concept of equilibrium charge, $\overline{Z}_a(\infty)$. The effective charge of a multicharge ion, Z_a^{eff}, is defined as the ratio between the actual values of the stopping power of a medium for the ions (under conditions of an equilibrium charge distribution) and the stopping power for protons having the same velocity:

$$Z_a^{\mathrm{eff}} = \sqrt{\left(-\frac{\mathrm{d}E}{\mathrm{d}x}\right)_a \Bigg/ \left(-\frac{\mathrm{d}E}{\mathrm{d}x}\right)_p}. \tag{4.12}$$

From this definition it is seen that Z_a^{eff} is the charge of a hypothetical charged particle that is always totally deprived of any electron shell, which would experience the same stopping in the medium as a real particle of charge Z_a.

Experiments show that the effective charge Z_a^{eff} may assume values within a broad range: it is very small at low ion velocities and tends toward the charge of a "bare" ion, Z_a, at high velocities (Fig. 4.7). Its dependence on the ion velocity can be approximated by

$$\frac{Z_a^{\mathrm{eff}}}{Z_a} \approx 1 - \exp(-v/v_B Z_a^{2/3}), \tag{4.13}$$

the physical meaning of which can be clarified by the following arguments. If the mean electron velocity inside an ion is much smaller than the velocity of the ion itself, the probability of stripping is high and the ion is not capable of carrying a large number of electrons with it. In the opposite case the process of the ion colliding with the atoms of the medium proceeds adiabatically, and electron stripping becomes difficult.

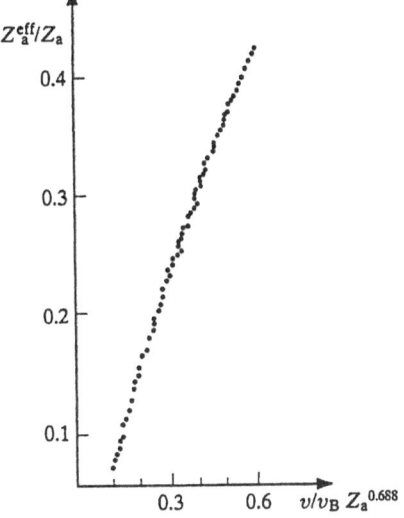

Fig. 4.7. Dependence of the effective ion charge on velocity in various media [4.6]

4.5 The Passage of Electrons Through Matter

The peculiarities of the passage of electrons through matter are directly due to their being light particles. First of all, the character of their ionization stopping and multiple scattering differs from what is known about, say, protons. Owing to the electron mass being small, large-angle scattering turns out to be so essential when electrons pass through matter that, unlike the case of heavy charged particles, an electron loses any memory of its initial direction of motion upon having undergone a certain number of collisions. For this reason, it has no sense to apply the usual concept of total path range in the case of electrons. Taking into account the confused trajectory of the electron motion, one must at least distiguish the mean path covered by the electron in a medium from its average distance (along a straight line) from the entrance point to the stopping medium (Fig. 4.8).

The second peculiarity, also due to the small electron mass, consists in ionization stopping having another mechanism added to it: namely, *radiative stopping* (bremsstrahlung):

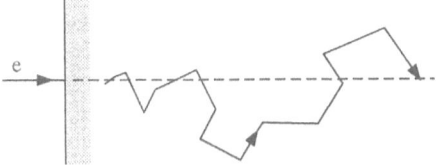

Fig. 4.8. Picture of an electron passing through a substance

$$-\frac{dE}{dt} = \frac{2e^2}{3c^3}|a|^2.$$

(4.14)

The acceleration a experienced by a charged particle in the Coulomb field of the target nuclei is inversely proportional to the particle mass:

$$a = \frac{1}{m}\frac{Z_1 Z_2 e^2}{r^2}.$$

(4.15)

Hence it is seen that the excess of the radiative stopping of electrons over the radiative stopping of protons is determined by the factor $(m_p/m_e)^2$, i. e., it amounts to over six orders of magnitude greater.

In quantum electrodynamics, electron bremsstrahlung is described by the diagrams

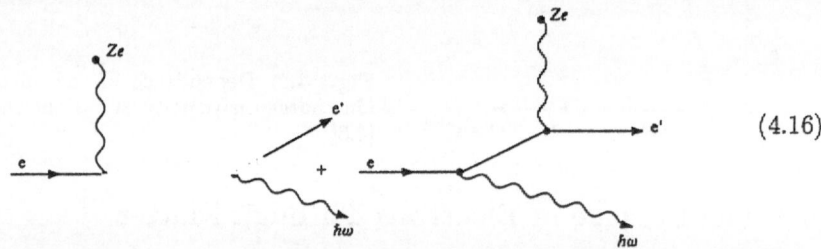

(4.16)

where the wavy line connected to the point Ze corresponds to the interaction (one-photon exchange) between the passing electron and the nucleus. The bremsstrahlung intensity is determined by the sum of the amplitudes (4.16) raised to power 2, i. e., it is proportional to the square charge of the nucleus, Z^2. At the same time the probability of bremsstrahlung in the case of collisions with individual atomic electrons (if they are considered free) is proportional to Z. For quantitative calculations it is essential that in (4.16) the electron interacts with a screened nucleus, instead of a "bare" nucleus; the screening effects are especially strong when the wavelength of the virtual photon is large as compared with the size of the atom. Naturally, the result of calculations depends on the atomic model chosen − on the form of the electron wave functions and the charge density distribution in the atomic shell. We present the result yielded by quantum electrodynamics when the Thomas–Fermi model is applied:

$$\left(-\frac{dE}{dx}\right)_{\text{rad}} \approx 4En_0\alpha r_e^2 Z^2 \ln\frac{183}{Z^{1/3}}.$$

(4.17)

Here, E is the total electron energy, $(-dE/dx)_{\text{rad}}$ is the energy lost by the electron per unit path due to radiation, n_0 is the density of the atoms in the medium, $\alpha = e^2/\hbar c$ is the fine-structure constant, $r_e = e^2/m_e c^2 = 2.8 \times 10^{-13}$ cm is the "classical electron radius".

Formula (4.17) shows that bremsstrahlung of electrons is proportional to their total energy:

$$\left(-\frac{dE}{dx}\right)_{rad} = \frac{1}{x_0}E. \tag{4.18}$$

Here, we have introduced the parameter x_0 given by the expression

$$\frac{1}{x_0} = 4n_0\alpha r_e^2 Z^2 \ln\frac{183}{Z^{1/3}}; \tag{4.19}$$

it has the dimensionality of length and is called the *radiation length*. The physical meaning of the radiation length becomes quite clear if (4.18) is solved under the assumption that bremsstrahlung is the only sort of energy loss occurring when an electron passes through matter:

$$E(x) = E_0 e^{-x/x_0}. \tag{4.20}$$

In (4.20), x_0 is the radiation length: the path length covered by an electron before its energy is reduced by a factor of e owing to radiation. Another equivalent characteristic of bremsstrahlung is the radiation thickness x_r related to the radiation length via the density of the medium, ρ:

$$x_r = \rho x_0 = \left(4N_A\alpha r_e^2\frac{Z^2}{A}\ln\frac{183}{Z^{1/3}}\right)^{-1} \tag{4.21}$$

(here, we have expressed the ratio ρ/n_0 through the mass number of the atoms of the medium, A, and the Avogadro number N_A). From (4.21) it is seen that the radiation thickness falls rapidly (roughly speaking, like Z^{-1}) in a transition from light to heavy substances.

Thus, we are dealing with two mechanisms of electron stopping in matter – ionization and radiative:

$$\left(-\frac{dE}{dx}\right) = \left(-\frac{dE}{dx}\right)_{ioniz} + \left(-\frac{dE}{dx}\right)_{rad}. \tag{4.22}$$

The character of their dependence on the electron energy differs from that of the dependence on the parameters of the medium, therefore the relationships between the contributions of these two mechanisms differ in different media. Within a broad range of energies up to 10 MeV, ionization losses are dominant. Further, when the ionization losses are near to those of the plateau, while the radiative losses rise in proportion to E, the picture changes to the opposite. Comparing the Bethe–Bloch formula (3.31) with formula (4.17) for radiative losses we see that in a transition to substances with greater Z the relationship between the ionization and radiative losses changes in favor of the latter; in heavy substances the dominance of radiative losses over ionization losses starts earlier, at lower energies, than in light substances. However, even in such media as lead ($Z = 82$) or uranium ($Z = 92$) this happens at energies of about 10 MeV, i.e., already in the relativistic region.

To characterize quantitatively transition from the mode in which the contribution of ionization stopping is dominant to the mode of dominant bremsstrahlung, the concept of the *critical energy* E_{crit} of an electron is introduced, at which these contributions are by definition equal to each other. The following formula is sufficient for a rough estimation of this quantity:

$$E_{crit} \approx \frac{800\,\text{MeV}}{Z}. \qquad (4.23)$$

Table 4.1 presents parameters characterizing the passage of electrons through some media.

Table 4.1. Radiation thickness x_r and critical electron energy E_{crit} for some substances

Matter	$x_r/\,\text{g}\,\text{cm}^{-2}$	$E_{crit}/\,\text{MeV}$
Air	37	80
Al	24	40
Pb	6	7.6

To conclude we shall schematically compare the main characteristics of the passage of protons and of electrons through matter (Fig. 4.9). The lower pair of pictures in Fig. 4.9 show how the particle beam intensity depends on the thickness of the layer of matter covered. In the case of an electron beam the distribution $N(x)$ exhibits an extended tail at large x; it is preceded by a nearly linear part of the curve. Continuing this linear segment downward we find at its intersection with the x axis a certain point R_{extrap}. The quantity R_{extrap} is conventionally used as a characteristic of the electron path in matter; this is the so-called electron *extrapolation path*. The following empirical relationship exists between the extrapolation path of an electron and its total energy:

$$R_{extrap}\,[\text{g/cm}^2] = 0.526E\,[\text{MeV}] - 0.094. \qquad (4.24)$$

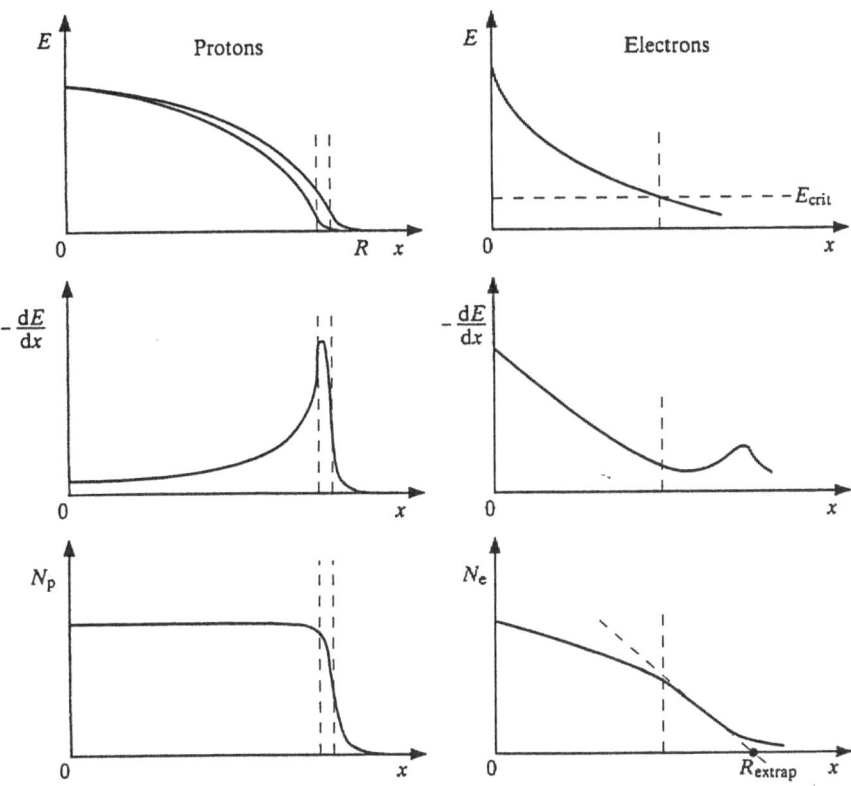

Fig. 4.9. Schematic plots presenting the main characteristics of the transmission of protons and electrons through matter

Fifth Lecture

5.1 Channelling

In this lecture we shall examine the peculiarities of the passage of fast charged particles through crystals. *Channelling* is one of the most striking effects indicating the ordered structure of a crystalline medium.

At the beginning of the century, Stark made the assumption that a positively charged particle entering a crystal along one of the directions of its atomic rows moves as if it were in a pipe, made up of the nearest of such rows, without approaching the atoms of the crystal. In modern investigations we make use of the term *channel*: we use *axial channel* if it is composed of several adjacent rows of atoms, or *plane channel* if it comprises two parallel crystallographic planes (Fig. 5.1). The phenomenon of channelling was first discovered in a "computer experiment", when the method of statistical simulation was applied in computing the passage of fast positively charged particles through a monocrystal (Fig. 5.2). The path range of a channelled particle is significantly larger than that of a particle passing through the crystal in an arbitrary direction, because the electron density and, consequently, the ionization stopping of particles in the middle of a channel is lower than the stopping averaged over the whole sample.

Enhancement of the path range or, accordingly, a decrease in the rate of ionization losses experienced by a channelled particle is not the only manifestation of channelling. A channelled particle collides less often with the atomic nuclei of a crystal, and this means that the probability of nuclear reactions occuring in a monocrystal when charged particles impinge upon it along cer-

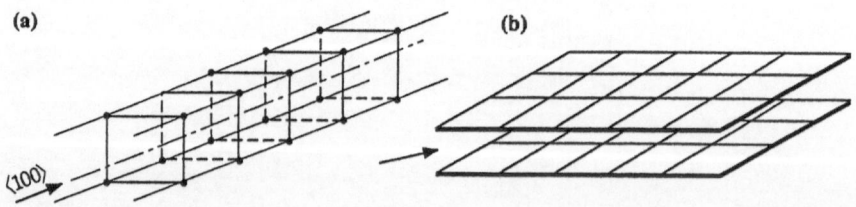

Fig. 5.1a, b. Schematic pictures of (a) an axial channel (100) in a simple cubic lattice and (b) a plane channel

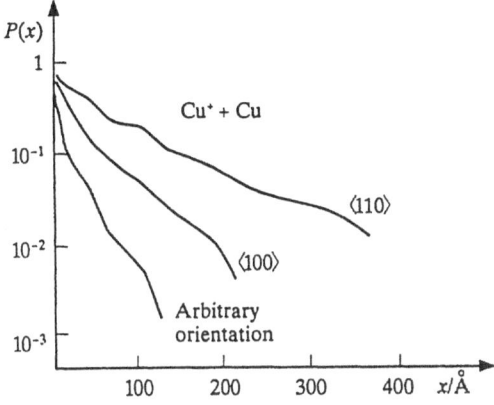

Fig. 5.2. Probability of Cu^+ ions with a 5 keV initial energy passing through a monocrystal of copper and losing energy down to 25 eV. Calculated by statistical simulation [5.1]

tain crystallographic directions is lower than when the particles travel in an arbitrary direction. For the same reason, a channelled particle gives rise to fewer defects in the lattice of the sample than a nonchannelled particle. Moreover, it interacts less often with the electrons of the inner atomic shells of the crystal than does a nonchannelled particle, so it gives rise to fewer vacancies in these inner shells. This, in turn, manifests itself in a weakening of the yield of characteristic X-ray or Auger radiation from atoms of the sample.

What are the conditions for confining a particle to a channel? First imagine a charged particle happening to be precisely in the middle of a channel and moving along its axis. In a perfect lattice at rest the forces exerted by the crystal on the particle in the transverse direction balance each other, so the particle should remain inside the channel and move along it at equal distances from the walls of the channel. The issue is not so simple in a real crystal. Let our particle carry a positive charge (such as does a proton or an α particle). Happening to be at a certain moment in time closer to one of the rows of atoms forming the channel, the particle begins to experience greater repulsion from this row than from the other ones. This repulsion tends to make the particle return to the central region of the channel, but whether the repulsion is sufficient to confine the particle to the channel depends on the strength and shape of the electrostatic force field acting on the particle inside the channel and on the "transverse" kinetic energy $E_\perp$ exhibited by the given particle.

We shall take advantage of the continuous potential approximation formulated by Lindhard in the 1960s, a time when the theoretical foundations of channelling were under development. With the aid of this approximation we shall introduce one of the most essential quantitative characteristics of the channelling phenomenon, the *limit channelling angle* (the "Lindhard angle").

Consider a fast charged particle moving at a distance ρ from a certain chain of atoms of a crystal (Fig. 5.3). Assuming the chain to be infinite, we

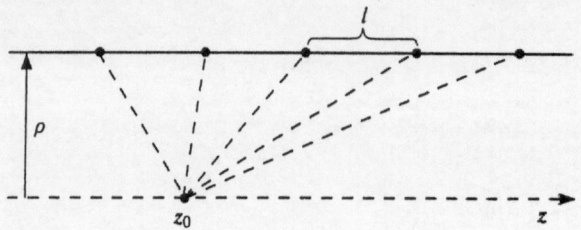

Fig. 5.3. Interaction of channeled particle with a chain of atoms in a crystal

calculate the total potential of the whole chain at point z_0 in the particle trajectory:

$$V(\rho, z_0) = \sum_n V\left[\sqrt{(z_0 - nl)^2 + \rho^2}\,\right], \tag{5.1}$$

where $V(r)$ is the potential in the interaction of the particle with a single atom. The simultaneous influence of a large number of atoms on the particle may be approximately described with the aid of the averaged potential

$$\overline{V}(\rho) = \frac{1}{l} \int V\left(\sqrt{z^2 + \rho^2}\right)\mathrm{d}z. \tag{5.2}$$

Precisely such a potential field is created around a continuous string, each element of which, $\mathrm{d}z$, interacts with the particle in accordance with $(\mathrm{d}z/l)V(r)$. Thus, in the continuous-potential approximation a discrete row of atoms is replaced by an infinite string.

By analogy with the potential of a string it is possible to introduce the averaged continuous potential of an infinite set of atoms lying in the given crystalline layer:

$$\overline{V}(h) = \frac{2\pi}{s} \int\limits_0^\infty \rho V\left(\sqrt{\rho^2 + h^2}\right)\mathrm{d}\rho, \tag{5.3}$$

where s is the area per atom and h is the distance of the particle from the layer. Two adjacent layers at a distance l from each other give rise to the following potential field in the plane channel:

$$V_c(x) = \overline{V}\left(\frac{1}{2}l - x\right) + \overline{V}\left(\frac{1}{2}l + x\right); \tag{5.4}$$

naturally, the concrete form of the field depends on the structure of the crystalline lattice and on the interaction potential between the particles and an individual atom; its general shape is shown in Fig. 5.4.

We shall now find the Lindhard angle for the case of plane channelling. In the continuous-potential model, the longitudinal motion of a particle and its transverse oscillations are independent of each other. According to this model, a particle moving along the channel does not feel the periodicity of the crystalline field, and the velocity of its longitudinal motion falls monotonically, owing to particles colliding with the electrons of the crystal. As to

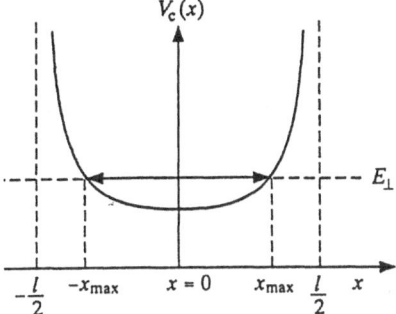

Fig. 5.4. Transverse oscillations of a positively charged particle in a plane channel

transverse oscillations, their amplitude x_{max} is greater, the higher the total energy of the transverse motion of the particles,

$$E_\perp = E \sin^2 \vartheta \,, \tag{5.5}$$

and is determined by the equation

$$V_c(x_{max}) - V_c(0) = E_\perp \,. \tag{5.6}$$

Here E is the total energy of the particles and ϑ is the angle of incidence of the particles upon the surface of the crystal, measured with respect to the axis of the channel (Fig. 5.5). Stable transverse oscillations of a particle in a one-dimensional potential well, $V_c(x)$, are possible only if their amplitude x_{max} is not too large. If, on the contrary, it is larger than a certain critical value x_{crit}, the particle leaves the channel either by "breaking down" its wall, owing to the energy of its transverse motion being too large, or by drastically changing the energy of the longitudinal motion as a result of coming near to some individual atom of the lattice (i. e., by undergoing large-angle scattering). The condition $x_{max} < x_{crit}$ leads to the inequality

$$E_\perp < V_c(x_{crit}) - V_c(0) \,, \tag{5.7}$$

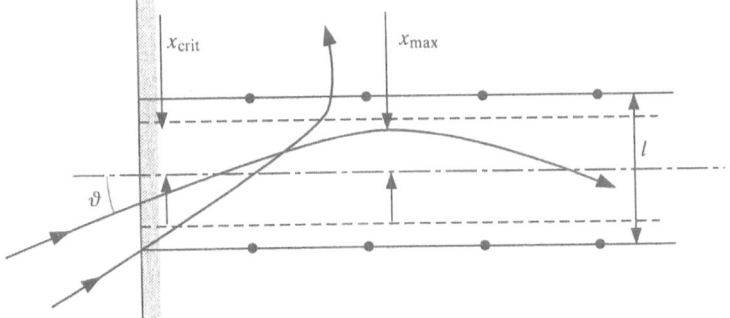

Fig. 5.5. Propagation of a positively charged particle in a plane channel

from which we obtain, taking (5.5) into account, the expression for the limit of the channelling angle ϑ_c:

$$\sin \vartheta < \sin \vartheta_c = \sqrt{\frac{V_c(x_{\text{crit}}) - V_c(0)}{E}}. \tag{5.8}$$

Naturally, this angle is smaller, the larger the energy of the particles entering the channel. The following simple formula is usually applied for a rough estimation of the limiting channelling angle:

$$\vartheta_c \approx \sqrt{\frac{2Z_a Z e^2}{Ed}}, \tag{5.9}$$

where d is the lattice parameter along the axis of the channel.

Thin crystalline films are used in atomic physics in the beam-foil method to test the theory of channelling. Using monochromatic ion beams, it is possible to measure with the aid of this method the energy losses of particles passing through very thin layers of matter, i. e., the actual stopping power of matter for particles of a given velocity. On the other hand, in those cases when a thin monocrystalline target can be oriented with high precision relative to the incident beam, such experiments yield valuable information on the channelling mechanism, on the electron density distribution over the profile of the channel (this problem is very important for solid-state physics), and on the influence of thermal oscillations of the lattice on the passing particle. Indeed, by varying the angle of rotation of the target with respect to the incident ion beam, the experimenter varies the mean impact parameter, that characterizes the passage of a channelled particle past the atoms of the lattice. One thus has the possibility of weakening or enhancing the contributions of individual electron shells of the atom in the crystal to the process of ionization stopping of the ion (we recall that when a charged particle moves in a disordered medium, no such selectivity exists: the contributions of different shells of atoms of the medium to the ionization losses of the particles "lose identity" when integration is performed over the impact parameter).

The picture of motion of a heavy charged particle in a channel is based on a combination of ideas from classical and quantum mechanics. Here, quantum mechanics merely plays an auxiliary role: it is used for interpreting the electron density distribution in a crystal, and issues of lattice dynamics can be dealt with, which is necessary, for instance, when the temperature dependence of channelling effects is studied. As to the motion of the channelled particle itself, as we shall see, its description is purely classical. What justifies this application of classical mechanics?

Let us take advantage of a concrete example in dealing with the issue (Fig. 5.6). Consider a He^+ ion moving along the $\langle 100 \rangle$ channel of a silicon monocrystal with an energy of 2 MeV. If the angle of deviation from the axis of the channel is set to be of the order of the Lindhard angle (here ϑ_c is of the order of 0.4°), then the energy of the transverse motion of the ion, $E_\perp$, will amount to approximately 2 MeV $\times (0.4°/57)^2 \approx 10^{-4}$ MeV. The

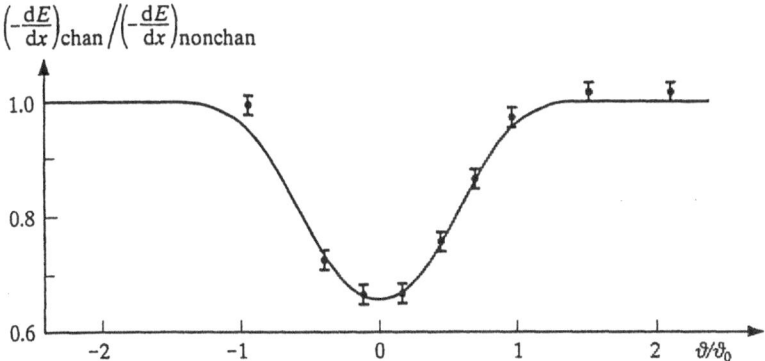

Fig. 5.6. Ratio between specific energy losses of channeled (in the axial channel $\langle 100 \rangle$) and nonchanneled 2 MeV He$^+$ ions transmitted through a thin film of a silicon monocrystal [5.2]. The *solid line* is a fit via the formula $(dE/dx)_{chan}/(dE/dx)_{nonchan} = 1 - \text{const} \cdot \exp(-2\vartheta^2/\vartheta_0^2)$ för $\vartheta_0 = 0.41 \pm 0.01°$

corresponding de Broglie wavelength $\lambda_\perp$ of transverse motion of the particle is $2\pi\hbar/\sqrt{2ME_\perp} \approx 10^{-10}$ cm. This is two orders of magnitude smaller than the width of the "well" in which an ion oscillates from one wall of the channel to the other. It can be seen from the above that the wavelength $\lambda_\perp$ is inversely proportional to the angle of deviation from the axis of the channel. This means that even for $\vartheta \sim 0.1\vartheta_c$ the de Broglie wavelength $\lambda_\perp$ still remains, in our case, an order of magnitude smaller than the distance between the walls of the channel. This just means that the particles exhibit transverse motion of a practically classical character.

However, classical mechanics cannot be applied if a light particle, say a positron, is made to enter the same channel. Quantum mechanics interprets the transverse oscillations of a channelled particle as the motion of a wave packet; its initial shape is determined by the conditions under which the particle enters the channel, whereas the character of its evolution can be investigated by solving the appropriate Schrödinger equation. The case of electron channelling requires separate consideration since, here, repulsion from the nodes of the crystalline lattice becomes attraction and the particle's motion in the channel turns out to be quite different from that of a positively charged particle. Its quantitative description requires quantum mechanics. From the point of view of classical mechanics, in the case of axial channelling, for instance, the electron moves along the channel as if it were wound around the chain of atoms of the crystal. We shall come back to the problem of electron channelling in Lecture 9.

Interesting quantum effects are revealed when ions totally deprived of electron shells undergo channelling. Imagine such an ion moving, say, along the axial channel of a certain crystal. Let us examine the scheme of its motion (Fig. 5.4). So far we have adopted the continuous-potential concept and assumed that the channelled particle does not to feel the periodicity of the crystalline field in its motion along the channel. Actually, this is not so. Along a significant part of its path it passes many nodes of the lattice at moments separated by strictly definite time periods $\tau = l/v$ and, thus, is subjected to a periodic perturbation from the lattice with an angular frequency of the main harmonic $\omega = 2\pi/\tau = 2\pi v/l$.

Generally, the periodic perturbation considered is weak and, when the velocity v of the ion is arbitrary, cannot exert any significant influence on it. However, for a definite velocity v the perturbation frequency ω may coincide with one of the eigenfrequencies of the ion, ω_{fi}, determined by the transition energy from its ground state $|i\rangle$ to a certain excited state $|f\rangle$. In conditions of such a resonance, when $\omega = \frac{1}{\hbar}(\varepsilon_f - \varepsilon_i)$, even a weak periodic perturbation can give rise to very strong transition of the ion from one state to another. Such an effect of resonance excitation of an ion moving along a channel was predicted by Okorokov and was named after him. The *Okorokov effect* has been observed in various experiments. One of the ways of observing it consists in the measurement of the energy dependence of the "survival fraction" of channelled ions, i.e., of the energy dependence of the fraction of ions that pass through the sample without changing their charge state. The point is that when resonance excitation takes place, the intrinsic sizes of ions increase drastically, as do their interaction cross sections of interactions with the crystal atoms and, for instance, the cross section of their losing electrons. This is

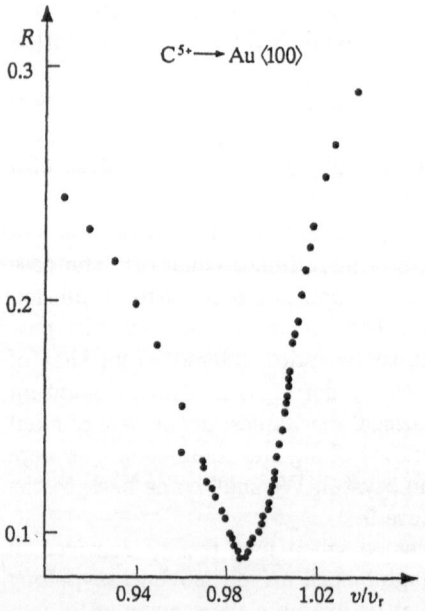

Fig. 5.7. Okorokov effect for C^{5+} ions transmitted along the $\langle 100 \rangle$ channel of a $1\,600\,\text{Å}$ thick crystal of gold [5.3]. The horizontal axis represents the ratio of the velocity of the ion particle v to the "resonance" velocity v_r corresponding to the $(n = 1) \Rightarrow (n = 2)$ transition (the main harmonic); in this case $E_r = 81.6\,\text{MeV}$. The vertical axis represents the fraction of C^{5+} ions transmitted through the sample without changing their charge state ("survival fraction")

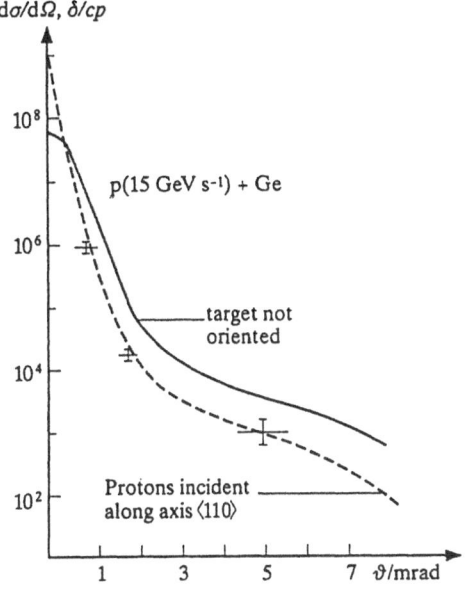

Fig. **5.8.** Angular dependence of the output of secondary charged particles in the interaction between relativistic protons and target nuclei in a germanium monocrystal [5.4]

manifest in the form of characteristic minima of the energy dependence curve of the survival fraction (Fig. 5.7).

To conclude the discussion of the channelling problem we stress that the energy range of charged particles in which this phenomenon is observed and studied is extremely wide – up to hundreds of GeV. The data presented in Fig. 5.8 as an example show the channelling effect in nuclear interactions of protons of momentum 15 GeV/c in a germanium monocrystal.

5.2 The Shadow Effect (Blocking)

Another effect, which, like channelling, is due to the arrangement of the atoms of a crystalline substance being ordered, is the *shadow effect* (blocking). It can be observed when charged particles recorded by detectors leave the lattice nodes of a monocrystal (Fig. 5.9); in this case, the Coulomb fields of the nearest atoms in the crystal block the directions of the principal axes and crystalline planes, so that in the angular distributions of particles leaving a monocrystal, regions of sharply reduced intensity, "shadows", form (Tulinov, 1965). The shadow effect was revealed in studies of nuclear reactions (including nuclear scattering of charged particles) on monocrystals and also in studies of the spontaneous α decay of heavy radioactive nuclei introduced into a monocrystal (Figs. 5.10, 5.11).

Figure 5.10 presents an example of the shadow effect for 3 MeV protons undergoing elastic scattering from a tungsten monocrystal. X-ray methods

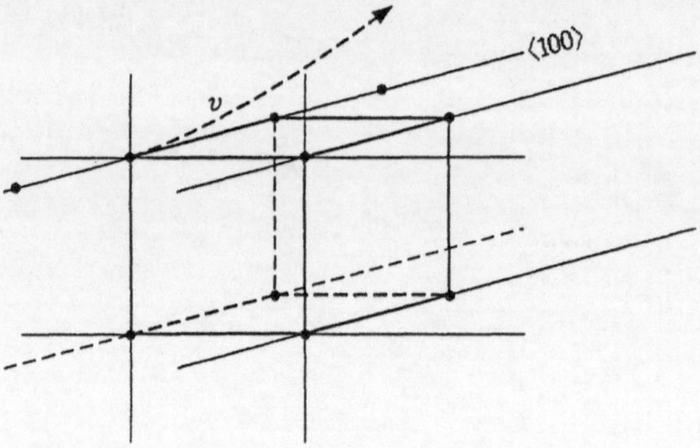

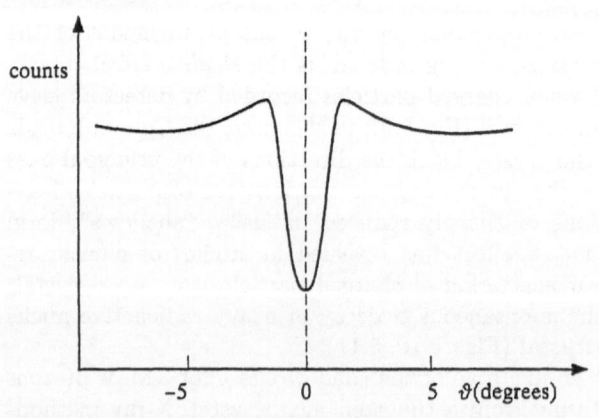

Fig. 5.9. Blocking of the ⟨100⟩ direction when a charged particle is emitted from a node of the crystalline lattice

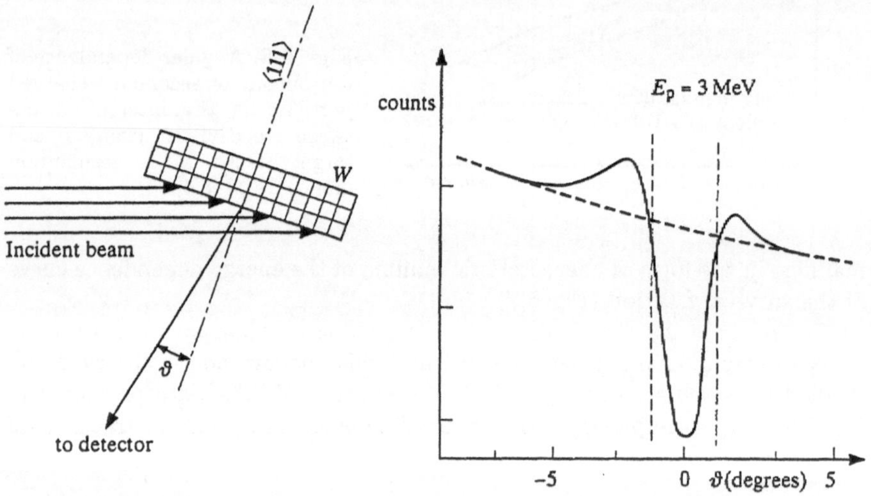

Fig. 5.10. Shadow effect in elastic scattering of 3 MeV protons from a tungsten monocrystal [5.5]

Fig. 5.11. Shadow effect due to emission of α particles by an ^{222}Rn isotope introduced into a tungsten monocrystal [5.6]; ϑ is the departure angle of the α particles relative to the ⟨111⟩ axis

have been applied in orienting the crystal with respect to the incident proton beam so that its most closely packed axis $\langle 111 \rangle$ was oriented at a large angle to the beam. The solid line in the picture follows the experimental points; the dashed line corresponds to the distribution of protons scattered by a poly-crystalline sample. The shape of the shadow is not trivial, it depends on many factors: the packing density along the axis of the crystal, the energy of the incident particles, and the charge of the particle produced in the scattering (to be more precise, its interaction with an invividual atom of the crystal). Investigations reveal that much information on the properties of crystalline substances can be obtained by studying the temperature dependence of the shape of the shadow.

Now, consider the elementary theory of how the shadow originates (Fig. 5.12). Let the particle source A and the scattering center B be at rest at a distance l from each other. The trajectory of a particle departing from the source at an angle ϑ_0 to the axis of the crystal can be found from the laws of classical mechanics, by taking advantage, for reasons of simplicity, of the approximation of small scattering angles. If the interaction potential of our particle with the scattering center B is approximated by the power function $V(r) = \alpha/r^n$ $(n > 0)$, then in this approximation the scattering angle of the particle, $\Delta\vartheta$, is expressed via the collision impact parameter b by the formula

$$\Delta\vartheta = \frac{2\pi\sqrt{\pi}}{m_a v^2 b^n} \frac{\Gamma((n+1)/2)}{\Gamma(n/2)}, \tag{5.10}$$

where v is the particle velocity. In our case $b \approx l\vartheta_0$ and, consequently, given the intial angle ϑ_0, we can determine the particle's departure angle ϑ from the expression

$$\vartheta = \vartheta_0 + \frac{2\pi\sqrt{\pi}}{m_a v^2 l^n} \frac{\Gamma((n+1)/2)}{\Gamma(n/2)} \frac{1}{\vartheta_0^n}. \tag{5.11}$$

Hence we find the angular dimensions of the shadow behind the scattering center. We present the expressions for the Coulomb potential $(n = 1)$ and the potential involving the inverse quadratic dependence on the distance $(n = 2)$:

$$V_1(r) = \frac{\alpha}{r} \Rightarrow \vartheta_{\min} = 2\sqrt{\frac{2\alpha}{m_a v^2 l}}; \tag{5.12}$$

$$V_2(r) = \frac{\alpha}{r^2} \Rightarrow \vartheta_{\min} = 2\sqrt[3]{\frac{2\pi\alpha}{m_a v^2 l^2}}.$$

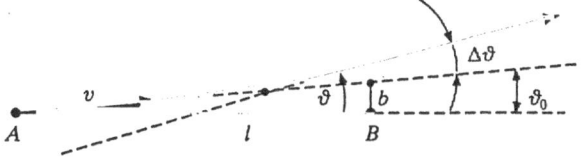

Fig. 5.12. Depiction of the origin of a shadow

In both cases the dimensions of the shadows are larger, the closer the source of particles is to the scattering center, i. e., the more closely packed the row of atoms oriented accordingly in the crystal is.

In a perfect lattice, each atom of the crystal is found at the intersection node of the crystallographic axes and planes., Owing to the translational symmetry of the crystal, the blocking effect arises along the same directions for each of the emitters occupying identical positions in the cells. Therefore, when charged particles leaving a monocrystal are registered with the aid of a detector covering a large solid angle (for instance, photographic plates), there arise characteristic "protonograms" (Fig. 5.13), on which dark spots and lines represent the directions of the crystal axes and planes.

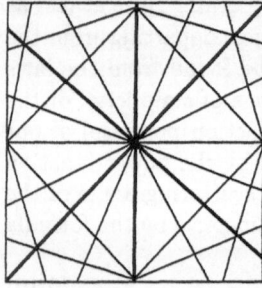

Fig. 5.13. Protonogram of a molybdenum crystal obtained by the scattering of 500 keV protons [5.7]. The central spot corresponds to the $\langle 100 \rangle$ axis, it is crossed by lines which are shadows of the planes (100), (110), and others

In our significantly simplified presentation we have set aside everything related to oscillations of the crystal lattice. These oscillations actually have an important influence on the blocking effect and on the shape of the shadow by determining its dependence on the temperature of the sample. At first sight, it may seem that heating the crystal and subsequent enhancement of the oscillation amplitudes of the atoms relative to each other should always result in a reduction in and narrowing of the darkened angular region. The mechanism of the influence of thermal oscillations on the shape of the shadow is actually much more complex, since the oscillations of atoms in the lattice are generally not chaotic and independent of each other, but represent inter-related oscillations involving strong correlations between the displacements of atoms both close to and far from each other.

Very often, channelling and blocking represent two aspects of one and the same phenomenon: the influence of the ordered crystal structure on the passage of charged particles through it. Thus, if the particles are produced inside the crystal, then their angular distribution at the exit from the sample exhibits features of either channelling or blocking, depending on the actual point in the crystal cell at which these particles are produced. Precisely in this way, from the angular distribution of μ^+ muons produced in the pion decay $\pi^+ \rightarrow \mu^+ + \nu_\mu$, it turned out to be possible, for example, to investigate the behavior of a π^+ pion (after its stopping) in the crystal lattice. Here, several

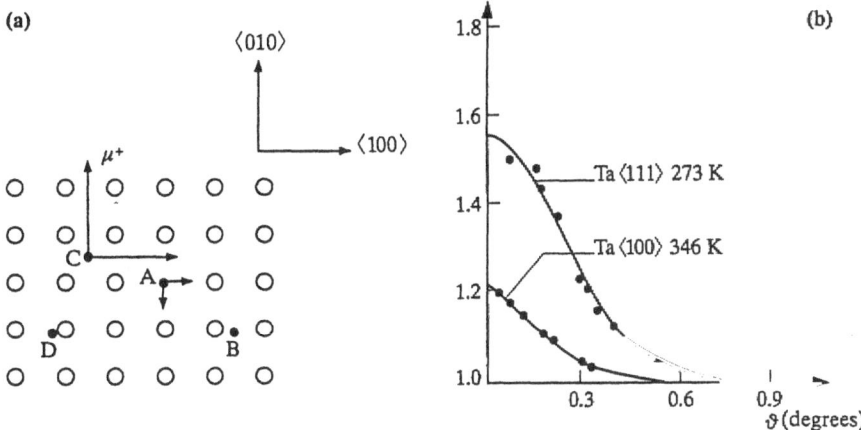

Fig. 5.14a, b. Channeling and blocking of μ^+ muons from the pion decay, $\pi^+ \rightarrow \mu^+ + \nu_\mu$, in a crystalline sample. (a) Schematic representation of versions of pion localization at the moment of decay. (b) Angular dependence of μ^+ muon yield in pion decay in a tantalum monocrystal [5.8]

possibilities can be imagined: the positively charged pion may, like a proton or deuteron, occupy a place in some quite definite cell; it may undergo diffusion over the entire sample, passing from cell to cell, or be localized near some impurity or defect; finally, the following phenomenon cannot be excluded: the π^+ may form a neutral pionium atom $(\pi^+ e^-)$, which, in turn, is capable of participating in chemical bonding with atoms in the medium. Various possible sites for the π^+ at the moment of decay are shown schematically in Fig. 5.14. In the case of site A the yield of μ^+ is blocked in the $\langle 100 \rangle$ and $\langle 010 \rangle$ directions; in the case of site B blocking takes place in the $\langle 100 \rangle$ direction and enhancement of the yield, due to channelling, occurs in the $\langle 010 \rangle$ direction; in the case of site C the yield of μ^+ is enhanced in both the $\langle 100 \rangle$ and the $\langle 010 \rangle$ direction. The site D indicates the case in which the produced pionium $(\pi^+ e^-)$ turns out to be chemically coupled to one of the atoms of the crystal.

   Experimental studies of the decay of π mesons in a crystal, in particular owing to its relation to the problem of hydrogen and heavy hydrogen in crystals, yield valuable information on the structure of solid-state bodies.

Sixth Lecture

6.1 Interaction of Molecular Ions with Matter

The physical processes involved in the interaction of molecular ions with matter give rise to significant interest not only from a purely scientific standpoint, but also from the point of view of applied studies; however, investigations of such processes have not been as extensive and profound as similar studies for atomic ions. Elaboration of the theory of such processes is at an early stage of development.

An essentially new phenomenon, encountered in the transition from processes involving atomic ions to processes with molecular ions, is dissociation. As a rule, the dissociation potentials of molecular ions amount to several electron-volts, so a fast molecular ion happening to traverse matter readily breaks up into its constituents. Therefore, in most of the technically interesting processes in the interaction of ions with matter, such as, for instance, sputtering of the surfaces of solid-state bodies, implantation, the formation of dislocations in crystals, and so on, a molecular ion acts upon the sample in the same manner as would the entire set of its constituent atomic ions acting independently of each other. We note, in this connection, that utilization of molecular beams in the physics of particle interactions with matter was initiated in relation to the search for methods of enhancing the flux densities of incident atomic particles. Subsequently, however, it was revealed that, under certain conditions, it is possible in investigations of the dissociation process due to the interaction of molecular ions with matter to obtain very valuable information on the molecular ion itself and on the properties of matter. We shall present several such examples taken from experimental and theoretical studies of recent years.

The "Coulomb Explosion" Phenomenon

The essence of the phenomenon and the methods for observing it consist in the following. A well-collimated beam of fast molecular ions is directed toward a thin (below or about 100 Å) film. The time of flight of the ion through the film is so small that the relative positions of the nuclei forming the molecule seem to be frozen during this entire period of time. The probability of the molecular ion being stripped of its electron cloud is so high, here, that upon passing through the film its nuclei turn out to be deprived of all or nearly all their electrons. Thus, the forces of Coulomb repulsion of the nuclei from each other turn out not to be compensated, and their action results in the nuclei of the molecular ion continuing the motion set down for them by the initial conditions and flying apart from each other in opposite (in the center-of-mass system) directions; the instantaneous loss of electrons occurring when a fast molecular ion penetrates the target actually results in the ion "exploding".

Now, let us consider the mechanism of Coulomb explosion in greater detail taking advantage of a concrete example. Consider a well-collimated monoen-ergetic beam of HeH$^+$ ions impinging with an energy $E = 3$ MeV upon a thin (~ 100 Å) carbon film. The parameters of the structure and of the internal motion of the HeH$^+$ ion (Fig. 6.1) are known from quantum chemistry: the dissociation energy (via the channel HeH$^+$ $\rightarrow$ He^0 + H$^+$) is $\varepsilon_D \approx 1.8$ eV; the equilibrium distance between the nuclei in the lowest ($v = 0$) vibrational state is $r_0 = 0.79$ Å; the oscillation frequency (the distance between the oscillatory levels $v = 1$ and $v = 0$) is $\hbar\omega = 0.4$ eV; the distance between the rotational levels is two orders of magnitude smaller.

In the case of the chosen energy, $E = 3$ MeV, the velocity of the HeH$^+$ ion amounts to about 1.1×10^9 cm/s, and, consequently, the time of flight across the film, t_d, is about 10^{-15} s. This is an order of magnitude smaller than the period of nuclear oscillations, $T_v = 2\pi/\omega = 1.5 \times 10^{-14}$ s. Let us estimate

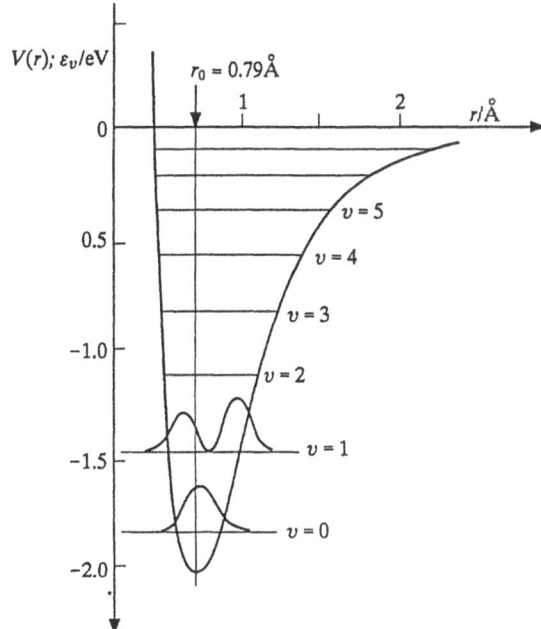

Fig. 6.1. Effective poten-tial energy of an interac-tion between the He and H nuclei in the HeH$^+$ molec-ular ion and the arrange-ment of its vibrational lev-els; the distribution den-sity of the distance be-tween the nuclei, $D_v(r) = |\psi_v(r)|^2$, is shown for the $v = 0$ and $v = 1$ levels [6.1]

the time required for an ion to lose its electrons after it has penetrated the target. The effective cross section of electron stripping, σ_{st}, is of the order of the geometric dimensions of an ion, i.e., about $10^{-16}\,cm^2$. In a medium of density ρ the average path length along which stripping takes place is $l_{st} = (\sigma_{st}\rho N_A)^{-1}$, where N_A is the Avogadro constant. This path corresponds to the time $t_{st} = (\sigma_{st}\rho N_A v)^{-1}$. Substitution of the appropriate figures yields $t_{st} \sim 10^{-17}$ s. Thus in the case under consideration the molecular ion loses its electrons almost instantaneously, if one takes into account that the transition time of an ion through a thin film amounts to $\sim 10^{-15}$ s and its vibration period to $\sim 10^{-14}$ s; we do not even mention the really slow rotational motion of the ion $(T_r \sim 10^{-12}$ s).

All possible orientations of the ion internuclear axis are equally probable in the incident beam. Now, let us follow the ion whose internuclear axis is at a certain angle ψ to the ion velocity vector at the moment of incidence upon the target (Fig. 6.2). Assume that after they have lost their electron shells, the helium and hydrogen (the He^{++} and H^+ ions) remain at a distance from each other equal to the equilibrium distance in the molecular ion, $r_0 = 0.79$ Å. Then, the Coulomb energy of their mutual repulsion is $\varepsilon_c = Z_1 Z_2 e^2/r_0 = 2e^2/r_0 = 36.5$ eV. Now, the kinetic energy of the nuclei flying apart is distributed between them (in the center-of-mass reference system) inversely proportionally to their masses: $4/5\varepsilon_c$ is transferred to the proton and $1/5\varepsilon_c$ to the α particle. How long does this take? Do the dissociation products He^{++} and H^+ have time to receive a dominant part of their share in the "explosion" energy before the entire $He^{++}-H^+$ complex passes through the film?

In the center-of-mass system the time picture of the nuclei flying apart is determined by the equation

$$\mu \ddot{r} = \frac{Z_1 Z_2 e^2}{r^2}, \qquad r(0) = r_0, \tag{6.1}$$

where $\mu = m_1 m_2/m_1 + m_2$ is the reduced mass of the nuclei. Upon solving this equation we see that given an initial ion energy of 3 MeV and film

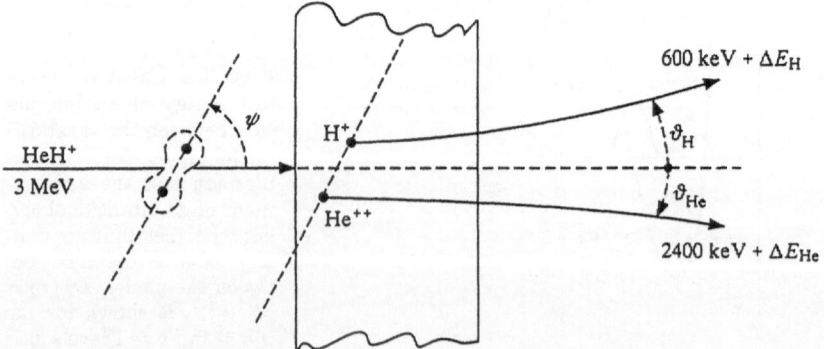

Fig. 6.2. Coulomb explosion occurring when a HeH^+ molecular ion is transmitted through a thin film

thickness of 100 Å the relative distance between the nuclei merely increases, during the transmission time through the film, from 0.79 Å up to 1.01 Å; i.e., the main part of the explosion energy (over 80%) is transformed into the kinetic energy of the nuclei only after they leave the film.

Now let us consider how the Coulomb explosion energy appears in measurements of the energies and angular distributions of the fragments. Let v_0 be the velocity vector of a molecular ion incident upon the film; it also represents the mean velocity vector of each of the nuclei composing the molecular ion before the "explosion". Let u be the velocity vector of a fragment in the center-of-mass system when the Coulomb explosion energy is fully realized. These two vectors differ very strongly in their absolute values; thus, in the example under consideration, the mean kinetic energy $E_0 = mv_0^2/2$ of a nucleus of the molecular ion before its dissociation and its kinetic energy $\varepsilon = mu^2/2$ in the center-of-mass system after the fragments separate amount to the following: 600 keV and 29.2 eV for the proton, and 2400 keV and 7.3 eV for the α particle. However, owing to summation of the vectors, $v_0 + u = v$, the energy shift of the fragment with respect to the mean value,

$$\Delta E = \frac{m(v_0 + u)^2}{2} - \frac{mv_0^2}{2} \approx mv_0 u \cos \psi \,, \tag{6.2}$$

and the change undergone by its motion in the laboratory system,

$$\vartheta \approx \frac{u}{v_0} \sin \psi \,, \tag{6.3}$$

turn out to be quite significant:

$$\vartheta_{max} = \vartheta \Big|_{\psi=90^\circ} = \frac{u}{v_0} = \sqrt{\frac{\varepsilon}{E_0}} \,, \tag{6.4}$$

$$\Delta E_{max} = \Delta E \Big|_{\psi=0^\circ;180^\circ} = \pm 2\sqrt{\varepsilon E_0} \,. \tag{6.5}$$

Thus, in the example considered we have

for protons: $\vartheta_{max} \approx 6.9 \text{ mrad } (0.4^\circ)$,
$\Delta E_{max} \approx \pm 8.4 \text{ keV}$;

for α particles: $\vartheta_{max} \approx 1.75 \text{ mrad}$,
$\Delta E_{max} \approx \pm 8.4 \text{ keV}$.

According to (6.2) and (6.3) the shift in a fragment's energy, ΔE, and the angle ϑ are related by the equation of an ellipse:

$$\left(\frac{\Delta E}{mv_0 u}\right)^2 + \left(\frac{\vartheta}{u/v_0}\right)^2 = 1 \,. \tag{6.6}$$

The numbers presented give an idea of the value of the effect in typical experimental conditions. However, in a certain very important aspect the computations performed above are too schematic: the distance between the nuclei in the molecular ion and, consequently, the initial distance $r(0)$ between them at the onset of the Coulomb explosion are not fixed, as assumed above, but are distributed in a certain manner about the mean value, depending on the relative populations of various vibrational states of the molecular ion in the incident beam. In each of such states the distance between the He^{++} and H^+ nuclei is distributed in accordance with the law

$$D_v(r) = |\psi_v(r)|^2, \qquad (6.7)$$

where $\psi_v(r)$ is the wave function of the vibrational motion of the nuclei in that state; the profiles of the distributions $D_v(r)$ for the states $v = 0$ and $v = 1$ of the HeH$^+$ molecular ion are shown in Fig. 6.1.

The vibration of nuclei in a molecule is manifest in the spread of the energy and angular distributions of the Coulomb explosion products. The ellipse (6.6) transforms into an elliptic ring, the thickness of which is greater the higher the amplitude of the nuclear oscillations. Figure 6.3 shows a typical "ring diagram" of the angular and energy distributions for the proton calculated in accordance with the scheme presented above and with the standard wave functions of the HeH$^+$ molecular ion for Coulomb explosion of the ion at an energy of 3 MeV.

The same distributions measured in experiments are shown in Fig. 6.4. The thickness of the film is greater here than the one assumed above, which means that the stage of fragments flying apart inside the target is more important. Comparison of the theoretical and experimental distributions of Figs. 6.3 and 6.4 reveals their unquestionable similarity, which concerns both the form of the ellipse (the average values of its semiaxes), and the mean thickness of the ring. In one aspect, however, these two distributions exhibit a sharp difference: in the experiment, when the outgoing proton travels along the incident ion beam ($\vartheta \to 0$), the symmetry of the energy distribution of the proton relative to the mean value $\Delta E = 0$ is strongly violated, the

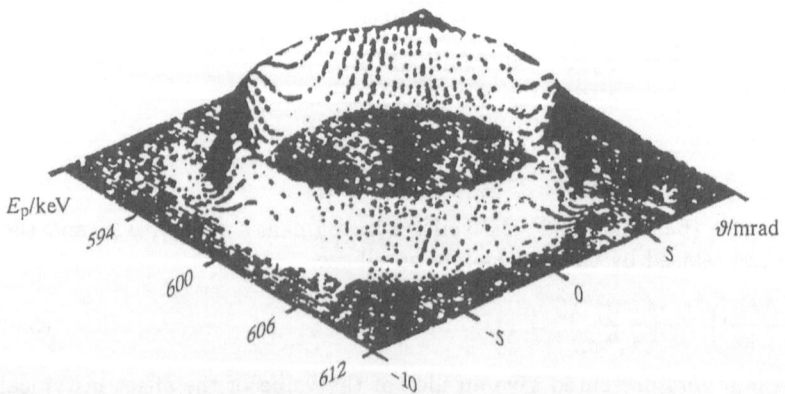

E_p/keV

594

600

606

612    $^-$10

ϑ/mrad

s

0

$^-$s

Fig. 6.3. Coulomb explosion of 3 MeV HeH$^+$ molecular ion passing through a thin film. The ring representing the angular and energy distributions has been calculated without taking into account the polarization of the medium

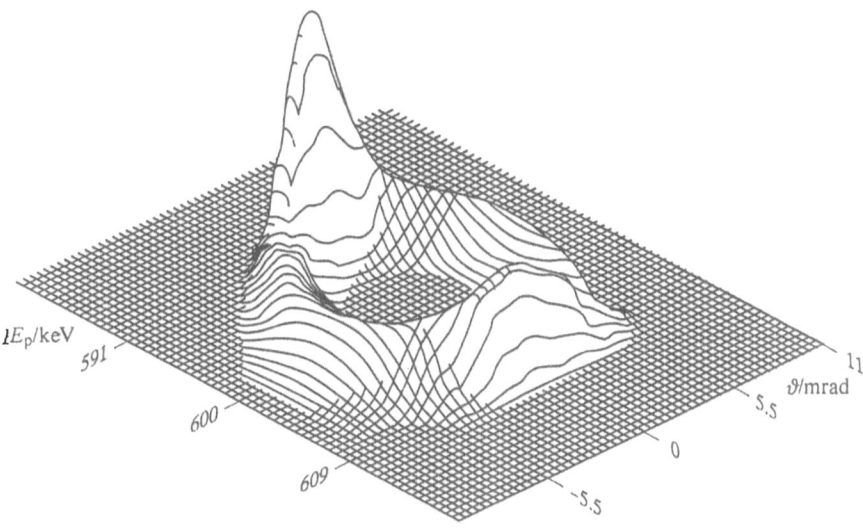

Fig. 6.4. The same distributions as in Fig. 6.3 – experimental results [6.2]; the carbon film is 195 Å thick

peak at the left ($\Delta E < 0$) of this distribution being sharply enhanced. This peak is due to protons that after dissociation of the molecular ion move immediately behind the foremost He^{++} ion and are, therefore, sensitive to the perturbations introduced by the ion into the medium. The foremost He^{++} ion polarizes the atoms of the medium and creates in it a certain "polarization wave", which, by analogy with a boat gliding on water, is also known as the *polarization wake*. Our proton happens to be precisely in the wake of this polarization track.

The notion of a polarization wave was introduced in the physics of particle interaction with matter back in the 1940s by Nils Bohr, but no exhaustive theoretical description of this phenomenon has ever been achieved. Observation of the wake effect in experiments devoted to studies of the Coulomb explosion has revealed that the dissociation products (the $He^{++}-H^+$ ion pair, in our case) of molecular ions travelling in a medium are not some system isolated from the medium, but together they form a unique entity – a certain complex *cluster* which, upon leaving the target, carries information not only on the molecular ion, but also on the dielectric properties of the medium.

We have taken advantage of the example of a two-atom molecular ion in considering the Coulomb explosion phenomenon. This phenomenon is also known to exist in the case of complex (multiatomic) molecular ions, and it is effectively applied in stereochemical studies of such ions. Here, correlation experiments involving the detection of two, or even three, dissociation products of the molecular ion in coincidence turn out to be especially informative.

The Reflection of Fast Molecular Ions from the Surface of a Crystal

In the course of studies the reflection of fast molecular ions from the surfaces of crystals, a phenomenon was observed (Molchanov, Soshka, 1965) which at first sight seemed strange: a significant fraction of the ions is reflected by a crystal without undergoing dissociation. Nondissociative reflection is observed even when the "transverse energy" transferred to the molecule from the target exceeds the dissociation energy of the molecule by orders of magnitude. This is what makes the phenomenon nontrivial. Indeed, the reflection of a molecule from a surface is actually the result of interaction between the atoms of the molecule and the atoms at the surface of the target. If we assume that in each individual collision the momentum transferred from the target to the molecule is taken up by a sole atom of the molecule, while the remaining atom or atoms of the molecule are merely "spectators" in this collision, then such a single collision should result in the bonds between the atoms in the molecule being broken. Since in a great number of cases this does not actually take place, it may be assumed that collisions between individual atoms of the molecule and the atoms of the target, occurring in the process of nondissociative reflection, are somehow correlated with each other (Fig. 6.5): although each of the atoms of the molecule receives a significant momentum, the perturbation introduced into their relative motion turns out to be small.

The theory of nondissociative reflection of molecular ions is still at the stage of development. Its most tangible success consists in the correct explanation of the dependence of the relative yield of molecular ions not having undergone dissociation (the "survival fraction") on the scattering angle (Fig. 6.6). At any rate, the hypothesis of the dominant role of correlated collisions in the process of nondissociative reflection has received strong support in special experiments performed with fast molecular ions and oriented monocrystals. A manifestation of the ordered arrangement of atoms throughout the entire volume of a crystal is the rows of atoms at its surface forming characteristic surface structures. In some directions (when the atoms of one row are centered between the atoms of adjacent rows) *semichannels* arise, in others (when ordered rows of atoms of different layers of the crystal happen to be strictly one under the other) deep *troughs* occur. Since the nondissociative reflection of molecular ions is due to correlated collisions with atoms

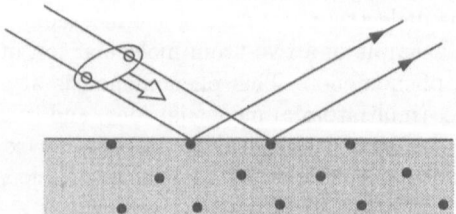

Fig. 6.5. Correlated atomic collisions in the case of nondissociative reflection of fast molecular ions from the surface of a solid

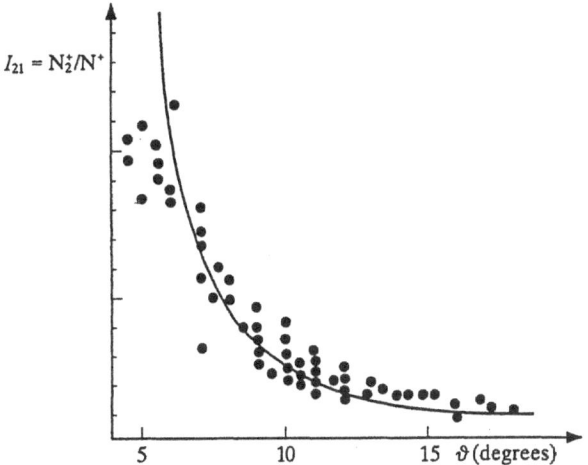

Fig. 6.6. Angular dependence of relative yield of N_2^+ ions ($E = 30\,\text{keV}$) in the case of reflection from the surface of polycrystalline copper; the curve corresponds to the θ^{-3} law for multiple scattering of the molecular ion from atoms of the crystalline surface [6.3]

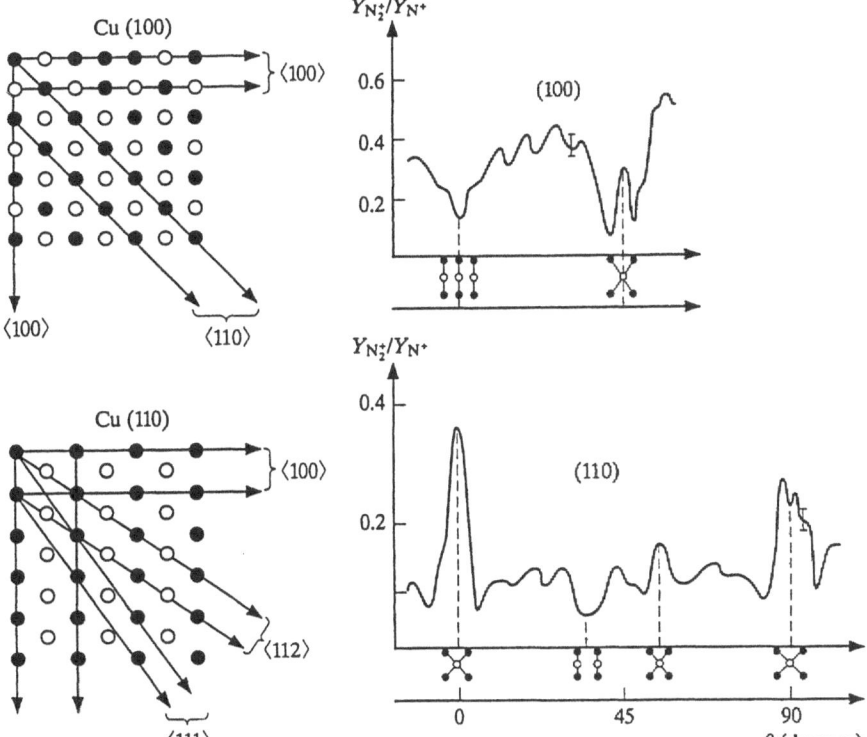

Fig. 6.7. Relative probability of nondissociative reflection of nitrogen molecular ions ($E = 30\,\text{keV}$) from (100) and (110) facets of a monocrystal of copper (for symmetric reflection: $\alpha = \frac{1}{2}\vartheta = 7°$) [6.4]. The *horizontal axis* is the azimuthal angle of the target counted from the $\langle 100 \rangle$ direction

of the target, it should depend on the mutual disposition of the atoms in the surface layer of the crystal and in the layers closest to it and, consequently, on how the single crystal is oriented with respect to the incident and reflected beams.

Experiments with monocrystalline targets have revealed a rigorous correlation between the presence at the surface of the target of some or of other ordered structures in the scattering plane, on the one hand, and the shape of singularities exhibited by the dependence of the survival fractions of molecular ions on the azimuthal rotation angle of the target, on the other hand: maxima always correspond to surface semichannels, and the minima of the curve correspond to troughs. By making a beam of molecular ions impinge upon different facets of a single crystal it is possible to observe along the same crystallographic axis either a maximum or a minimum in nondissociative reflection. Everything depends on what corresponds to the given axis on the given facet: a semichannel or a trough (Fig. 6.7).

Formation of Negative Molecular Ions in the Vicinity of the Surface of a Metal

Experimental studies reveal the high probability with which the interaction of molecules with the surface of a metal results in the formation of negative molecular ions. The energy of the affinity of the electron to a neutral molecule, ϵ, is always very small or at any rate lower than the work function for an electron emitted from the surface of the metal. Which mechanism, nevertheless, makes advantageous the capture by a molecule of an electron emitted from the surface? At present, this question gives rise to very considerable interest. In particular this is due to the results (the formation of negative molecular ions) of recent studies being perceived as a most important intermediate stage in many other processes occurring in the course of molecules interacting with the surface of a metal, such as their dissociation, chemisorption (the formation of chemical complexes at the surface), and strong vibrational excitation of the molecules. In connection with elaboration of this issue, a new concept termed the "harpoon mechanism" has appeared in the literature on the physics of molecular beams. Let us proceed to clarify its essence.

Consider a neutral molecule A_2 in a vacuum at a considerable distance from the surface of a metal. The energy spectrum of free electrons in the metal is presented schematically in Fig. 6.8. The shaded part of the diagram shows the part of the conductivity zone occupied by electrons, ε_F is the Fermi level. We shall consider our molecule and the metal to be a unified physical system. The level A_2 at the right of Fig. 6.8 corresponds to the state of a neutral molecule. When the molecule is far from the surface, the level A_2^- is raised over the level A_2 by $\Delta E = \varphi - \varepsilon$, where φ is the work function required for the emission of an electron from the surface of the metal into the vacuum.

Now, consider separately the interactions of the neutral molecule A_2 and of the negative ion A_2^- with the metal as they approach its surface without, for

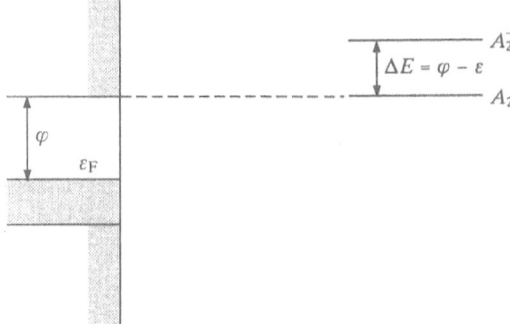

Fig. 6.8. Relative positions of energy levels in a molecule A_2 consisting of two atoms and in a molecular ion A_2^- when an electron is captured from the surface of the metal

the time being, taking into account possible transition of the electron from the metal to the molecule and back. In the first case, this interaction is "switched on" only in the very close vicinity of the metal, in which case it is repulsive. A totally different picture arises when the negative ion A_2^- approaches the metal. Polarization of the free electron gas in the metal by the negative charge of the A_2^- ion produces a positive image charge (symmetrically located under the surface of the metal), which results in the A_2^- ion being more and more attracted to the surface of the metal by this induced charge. Only at a very small distance, when the forces of interatomic interaction become felt, is this attraction replaced by repulsion. The curves of the potential interaction energies ("terms") for the neutral molecule A_2 and the negative ion A_2^- are shown schematically in Fig. 6.9; the horizontal axis represents the distance from the surface of the metal to the center of mass of the molecule A_2 and the ion A_2^-. We shall now examine the system composed of these two terms.

Figure 6.10 corresponds to the case when the initial kinetic energy E of the neutral molecule A_2 moving perpendicularly to the surface of the metal

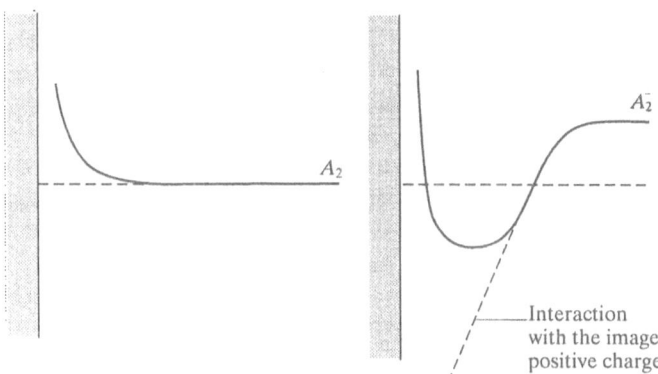

Fig. 6.9. Schematic picture of the interaction of a neutral molecule A_2 and of a molecular ion A_2^- with the surface of a metal, without the charge exchange process being taken into account

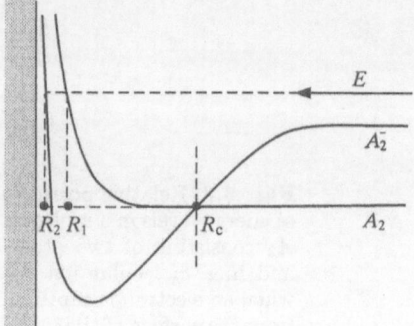

Fig. 6.10. Intersection of the terms of a neutral molecule A_2 and of a negative molecular ion A_2^- in the vicinity of a metallic surface

is higher than the transition energy of $A_2 \rightarrow A_2^-$: $E > \Delta E = \varphi - \varepsilon$. Until we take into account the possibility of such a transition, the motion of the molecule A_2 seems to be extremely simple: the molecule arrives at the classical turning point R_1, where all its kinetic energy is spent on overcoming the potential energy of repulsion from the surface, and then it repeats the same path in the opposite direction and proceeds toward infinity. We shall now take into account the possbility of the molecule capturing an electron from the surface of the metal. In principle, this can take place at any point on its trajectory; however, the probability of such a transition is generally very small, since everywhere, except in the vicinity of the intersection point of the terms R_c, a change in the electron state of the molecule, $A_2 - A_2^-$, requires an accordingly sharp change to occur in the kinetic energy of its forward motion. Only in the vicinity of the point R_c is transition of an electron from the metal to the molecule possible with a high probability. This means that a significant number of the A_2 molecules moving toward the surface of the metal transforms upon covering the distance R_c, into negative molecular ions A_2^-. Continuing their motion further toward the surface they reach the turning point R_2 and turn back. When again passing the intersection point R_c both the fraction of neutral A_2 molecules and the fraction of negative ions are again capable of exchanging an electron with the metal in both directions: $A_2 \rightleftharpoons A_2^-$. These processes result in the fraction of negative ions, A_2^-, present in a molecular beam reflected from the surface of a metal, being large.

In the above exposition there is an extremely essential point: the transition of an electron from the metal to the molecule takes place at a large distance from the surface of the metal. Hence the term *harpoon mechanism*: the light particle, the electron, "ejected" by the metal toward the approaching molecule plays the part of a harpoon with the aid of which this molecule (which becomes a negative ion) is strongly attracted to the surface.

The harpoon mechanism makes it easy to understand the important role played by the relationship between electronic and vibrational degrees of freedom of a molecule in dissociation processes due to its reflection from a surface. Figure 6.11 schematically presents the curves of the effective potential ener-

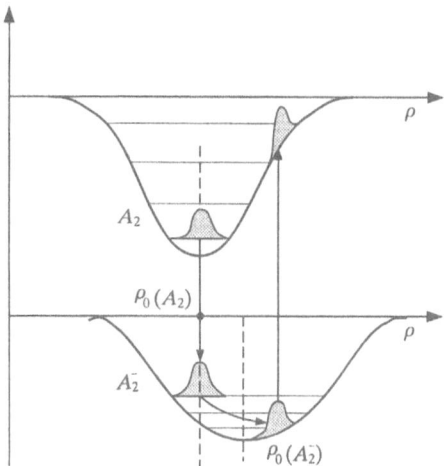

Fig. 6.11. Frank–Condon principle in connection with the $A_2 \rightleftharpoons A_2^-$ interaction in the vicinity of a metallic surface

gies, $U_{A_2}(\rho)$ and $U_{A_2^-}(\rho)$, for the internuclear interactions in the molecule A_2 and in the negative ion A_2^-. Here (not to be confused with Fig. 6.10!), the horizontal axis represents the distance ρ between the nuclei making up the molecule or the molecular ion. The mean equilibrium internuclear distances in the neutral molecule and in the negative ion are not the same: $\rho_0(A_2) \neq \rho_0(A_2^-)$. Consider a flux of neutral A_2 molecules incident upon the surface of a metal, and let the molecules not be excited vibrationally. Vibrations of the nuclei in the ground state take place about the equilibrium distance $\rho_0(A_2)$. As the molecules move forward toward the metal, somewhere at a distance R_c from the surface some of the A_2 molecules transform into negative ions A_2^-. When this takes place, the nuclei of the molecules, being heavy particles, are not in time to follow the transition of the electrons (the Frank–Condon principle; see Fig. C.11), so when they land in a new potential well $U_{A_2^-}(\rho)$, they remain as before at a mutual distance of about $\rho_0(A_2)$, which does not coincide with the equilibrium distance between them and the new well. Vibrations of the nuclei about this new equilibrium position $\rho_0(A_2^-)$ arise; in other words, an inherent feature of the process of stripping the electron from the metal to join the molecule consists in the vibrational states of the produced molecular ion A_2^- becoming populated intensively. The same takes place after reflection of the molecular ions from the surface of the metal, when they again pass the vicinity of the intersection point of the terms R_c. The result is a high probability for both the molecular ions and the neutral molecules leaving the metal for the vacuum being in excited vibrational states; i. e., they are predisposed to dissociation into their constituent atoms and atomic ions.

We have dealt with the case in which the kinetic energy of the forward motion of the molecules toward the surface of a metal (the "transverse" energy of the molecules) exceeds the $A_2 \rightarrow A_2^-$ transition energy. In the opposite

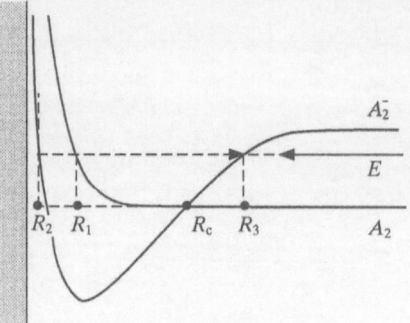

Fig. 6.12. "Harpoon mechanism" for the formation of negative molecular ions A_2^- in the vicinity of a metallic surface in the case of small "transverse" energies for the A_2 molecule

case, when $E < \Delta E = \varphi - \varepsilon$, no formation of free negative molecular ions A_2^-, capable of ultimately leaving the metal for a vacuum, is possible. However, according to the idea of the harpoon mechanism, negative molecular ions are produced intensively at the intermediate stage of reflection, dissociation, and chemisorption processes of neutral molecules (Fig. 6.12). In this case, the negative ions A_2^-, produced owing to the transition of electrons from the metal to the molecules, turn out to be "confined" within a limited region of motion between the turning points R_2 and R_3. From the above it is clear that their repeatedly crossing the vicinity of point R_c favors the intense population of excited vibrational states.

The whole exposition above concerns the case of molecules and molecular ions with relatively small velocities when the immediate impact of the incident heavy particles upon the target atoms plays no significant role in changing the internal state of the molecule. Here we speak of the transverse velocity component. If the molecular ions undergo grazing reflection from the surface of a crystal, then this normal component may be small, even if the initial kinetic energy of the incident particles is high. In this connection the

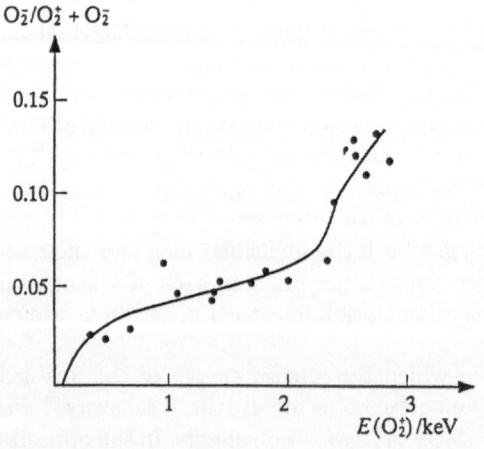

Fig. 6.13. Relative yield of negative oxygen molecular ions, O_2^-, in the case of grazing symmetric ($\alpha = 5°$) reflection of O_2^+ ions from the (111) facet of a crystal of silver [6.5]

harpoon mechanism concept is applied not only in the case of processes involving slow molecular ions, but also for explaining the high yield of negative molecular ions due to reflection of fast positively charged molecular ions from the surface of a crystal (Fig. 6.13).

6.2 Application of the Method of Computer Simulation

The method of computer simulation has deeply penetrated the practice of scientific investigations in the physics of particle interaction with matter. In the previous lecture, one example of such an application was mentioned: it was precisely a "computer experiment" that initiated modern studies of the channelling phenomenon (see Sect. 5.1). In this lecture we shall complement it with other examples. In doing so we shall set aside the mathematical aspects of the method and the statistical reliability of the results thus obtained, and concentrate on the physical essence of the actual processes being modelled.

A typical computation by the method of computer simulation in the field of particle interactions with matter starts with defining the interaction law of a single particle penetrating a sample with its elements (atoms); here, the properties of the target medium itself are considered to be known. All the problems to be dealt with are immediately separated into two sorts, two classes. The first includes those cases in which the perturbations introduced into the target medium by each individual particle of the incident flux turn out to be insignificant for other particles of this flux. In such cases it suffices for computer simulation of all possible events with the aid of standard Monte Carlo codes to follow each of the particles in the incident flux independently of the other particles. Problems of the second sort are much more difficult because it is necessary to take into account how changes occurring in the target accumulate as more and more particles impinge upon the target, or how the perturbation, even if instable, introduced by one particle in the target medium influences the next, follow-up particle.

Computer simulation of the processes in the interaction of charged particles with matter is based on the laws of both classical mechanics and quantum mechanics. Quantum mechanics is involved in these computations together with the ion–atom interaction potentials and the electron density distribution (within the limits of a single atom or of the entire sample); thermal oscillations of the atoms of a lattice in a solid and alignment phenomena of the angular momentum of the target atoms or of the incident ion etc. are interpreted at a quantum-mechanical level. Particle trajectories are derived from the equations of classical mechanics, given the initial conditions. The method of computer simulation is extremely illustrative. However, here also lurks the danger of overestimating the significance of this method, if one forgets that each time, before turning to the computer, we ourselves fix the actual *physical model* for the process of interest; i.e., already at this stage we simplify, by simulating, the actual picture of the interaction.

Simulation of the Reflection of Low-Energy Ions from the Surface of a Crystal

The reflection of charged particles from the surface of solids (*back scattering*) is widely applied for determining a whole number of parameters characterizing the properties of solids. The lighter the bombarding particles are and the higher their energy, the deeper they penetrate into the sample. Therefore, of particular interest is the reflection process of heavy ions of low energies, which carry information on a very small number of atomic layers in the target. The best known example of the application of this method is the investigation of a surface composition. The idea underlying this application is simple: depending on whether the incident ion undergoes collision with an atom of one or another component of the target surface it loses different fractions of its energy when scattered at a given angle. If the incident ion beam is monoenergetic, then the energy spectrum of ions scattered at a given angle will be expected to display separate lines, the position of each of which will indicate the mass of the atom with which the collision took place.

Indeed, consider a beam of ions of mass M_1 and of kinetic energy E_0 incident upon a target. Let ϑ be the scattering angle of an ion undergoing collision with the atom of mass M_2 at the surface of the target. The kinetic energy of the scattered ion is determined on the basis of the conservation laws for the particle energy and momentum. The calculations themselves are elementary, but the assumptions within the framework within which these calculations are actually performed deserve to be discussed. Suppose the target atom can be considered as free (i. e., its relationship with the surroundings may be neglected) and at rest (i. e., its thermal oscillations may be neglected). Then if $M_1 < M_2$, we will find the energy of the scattered particle to be

$$E_1(\vartheta) = E_0 \left(\frac{M_1}{M_1 + M_2} \right)^2 \left(\cos\vartheta + \sqrt{\frac{M_2^2}{M_1^2} - \sin^2\vartheta} \right)^2 . \tag{6.8}$$

Otherwise ($M_1 > M_2$) the scattering angle of the outgoing ion is limited by the value

$$\vartheta_{\max} = \arcsin \frac{M_2}{M_1} , \tag{6.9}$$

and within the cone $\vartheta < \vartheta_{\max}$ the $E_1(\vartheta)$ is two-valued: the sign before the radical in (6.8), in this case, is $\pm$.

Each mass value of the atoms at the target surface, M_2, M_2', etc., has a certain energy $E_1(\vartheta)$ corresponding to it. However, for a quantitative analysis of the target composition it is not sufficient to know the positions of the peaks in the energy spectrum of the scattered ions: the relative amounts (concentrations) of atoms of different sorts can be found only if the relative intensities of these peaks are established. These depend both on the kinematics of the elementary collision of an ion with a target atom and on many other factors. In analytical studies it is especially difficult to deal with multiple scattering

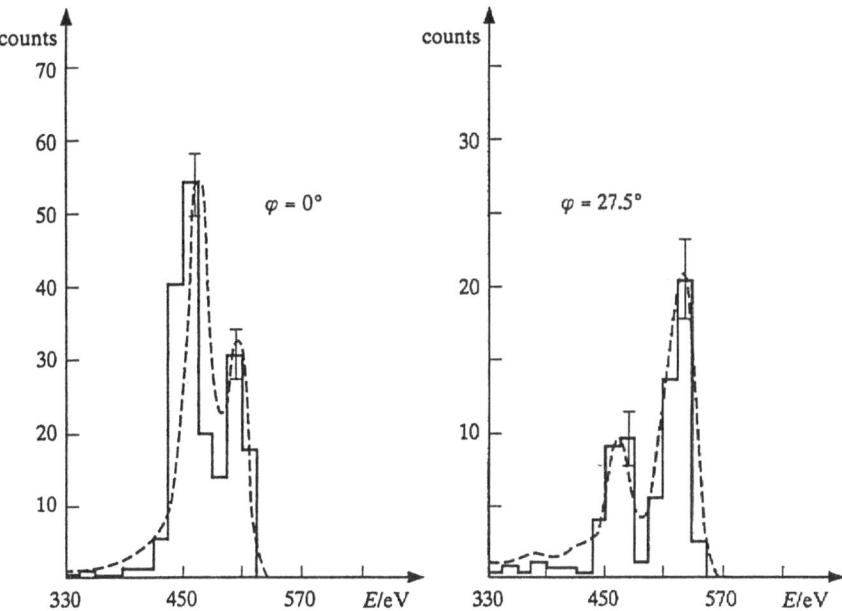

Fig. 6.14. Energy distribution of 600 eV K$^+$ ions reflected from the (100) surface of a monocrystal of gold; the structure of the surface layer is (1×2). The *dotted lines* represent experimental data; the histograms are the result of computer simulation. The angle φ is counted from the $\langle 100 \rangle$ direction [6.6]

effects, including those related to deep penetration of the scattered particles into the sample. It is in this case that computer simulation turns out to be indispensable, enabling one to follow the most intricate trajectories and to take into account their contributions to spectra and angular distributions of the reflected particles.

Let us examine a concrete example. Consider potassium ions K$^+$ of kinetic energy 600 eV impinging upon the (100) surface of a monocrystal of gold. Gold has a face-centered cubic lattice and is found to be in two modifications differing from each other in the structure of the (100) surface layer: (1×1) and (1×2). In the (1×2) version every other row of atoms of the surface layer in the $\langle 110 \rangle$ direction is absent, i. e., the lattice of the surface layer seems to be thinned out as compared with the "usual" (1×1) version. Figure 6.14 shows a spectrum of K$^+$ ions that have undergone nearly symmetric reflection from the surface of the sample; the angle of incidence is $\psi = 30°$, the scattering angle is $\vartheta = 70°$. In one case the target was oriented so the $\langle 110 \rangle$ axis at its surface was in the scattering plane (the azimuthal angle is $\varphi = 0°$), in the other case the target was rotated through an angle of $\varphi = 27.5°$. In both cases, instead of a single peak, which is to be expected if the whole process reduces to single scattering, a much more complex picture was seen with two large peaks in the spectrum of reflected ions. The position of the left-hand peak is quite in accordance with (6.8). The nature of the right-hand peak is revealed by computer simulation of the reflection process.

The histograms presented in Fig. 6.14 are obtained by summation of the different types of particle trajectories computed by the Monte Carlo method under the assumptions of single and of successive multiple collisions with the atoms of several upper layers of the crystal. On the whole, the calculation correctly reveals the shape of the spectrum and the character of its changes due to variation of the target orientation. The ion reflection mechanism is far from simple; this can be seen from Table 6.1. Nearly half of all the events (43 % for $\varphi = 0°$ and 48 % for $\varphi = 27.5°$) involve "zigzag" trajectories passing through a large number of layers of the crystal; they are the ones that in the main are responsible for the right-hand peak in the spectrum. Single collisions are mainly responsible for the left-hand peak. The relationship between the contributions of the uppermost and the next layers of the crystal varies strongly with rotation of the target.

Table 6.1. Computer simulation of reflection of 600 eV K^+ ions from the (100) (1×2) surface of a monocrystal of gold; $\varphi = 35°$; $\vartheta = 70°$ [6.6]

Class of trajectories	Relative contribution of trajectories (%) at different angles between scattering plane and direction $\langle 110 \rangle$	
	$\varphi = 0°$	$\varphi = 27.5°$
Single collisions with atoms of the first layer	5	11
Single collisions with atoms of the second or third layers	29	13
Zigzag trajectories involving atoms of the first two layers	7	11
Zigzag trajectories involving a large number of layers of the crystal	43	48
Double and multiple collisions without zigzags in the particle trajectories	16	17

Here we shall make a small, but important for the following, deviation. We see from the example in Fig. 6.14 that when a particle undergoes scattering at a certain angle, of all the collisions of different multiplicities it is single collisions that result in the particle losing maximum energy. This fact is due to a very general regularity inherent in the scattering of fast particles on a composite target, which is known not only in connection with the interaction of charged particles with matter, but also in other branches of physics:

Fig. 6.15. Scheme of double scattering of ions from a crystalline surface

in the case of neutron stopping, in the physics of nuclear reactions (where each individual nucleus represents a composite target), in high-energy hadron physics, etc. In this connection, we shall consider the energy losses of particles undergoing multiple scattering from a more general point of view, although schematically, and compare them with the losses occurring in the case of single scattering. Consider a particle of mass M_1 interacting with a pair of particles of mass M_2 to be scattered at a small angle $\vartheta \ll 1$ (Fig. 6.15). In the case of single scattering the energy lost by the particle $\Delta E(\vartheta) = E_0 - E_1(\vartheta)$ is calculated by the formula

$$\Delta E^{(1)} \approx \frac{M_1}{M_2} \vartheta^2 \cdot E_0 \,. \tag{6.10}$$

To consider the case of double collision we assume our pair of particles M_2 to be in the scattering plane. Then, deviation by an angle ϑ can be represented, in the case of two collisions, as scattering at angles ϑ_1 and ϑ_2 from the first and second particles of mass M_2, and $\vartheta_1 + \vartheta_2 = \vartheta$. In this case, unlike in (6.10), the energy loss,

$$\Delta E^{(2)}(\vartheta_1, \vartheta_2) = \frac{M_1}{M_2}(\vartheta_1^2 + \vartheta_2^2)E_0 \,, \tag{6.11}$$

is not determined uniquely by the scattering angle ϑ. Substitution of $\vartheta_1 = \vartheta_2 = {}^1/_2\vartheta$ into (6.11) yields the lost energy, which is two times smaller than in the case of single scattering:

$$\Delta E^{(2)}({}^1/_2\vartheta, {}^1/_2\vartheta) = {}^1/_2\Delta E^{(1)}(\vartheta) \,. \tag{6.12}$$

In the case of two collisions the entire spectrum of lost energy occupies the interval from ${}^1/_2\Delta E^{(1)}(\vartheta)$ to $\Delta E^{(1)}(\vartheta)$, but the shape of this spectrum is no longer determined only by the kinematics of the process, but is now also determined by the angular distribution of particles scattered in the elementary collision (by the dependence of the scattering probability on the impact parameter), i. e., by the dynamics of the process of interaction between the particles with masses M_1 and M_2; it also depends on the mutual disposition of the particles with mass M_2.

We shall now go on discussing the results of a computer simulation of the reflection of low-energy ions from the surface of a crystal. We saw that multiple collisions contribute greatly to the main characteristics of the reflection process – the angular and energy distributions of the reflected ions. Hence, the shape of these distributions should be sensitive to parameters characterizing the mutual disposition of the nearest neighbors in the lattice of a crystal. This circumstance has made possible, with the aid of the method of back scattering of low-energy ions, the solution of a number of important diagnostic problems of the surfaces of cyrstals. We shall mention some of these problems related to the properties of pure surfaces: observation of phase transition from the structure of the surface layer of sort (1×1) to the (1×2) structure in single crystals of platinum and gold; high-precision measurement of the distance between the first and second surface layers (for molybdenum, for instance, it is approximately 0.2 Å smaller than the corresponding distance between two adjacent layers throughout the whole volume of the crystal); investigation of correlations of the nearest order at the surface of single crystals and, in particular, of their stability when the phase transition point is reached. In all these cases, data on the parameters of the crystals were derived with the aid of computer simulation.

Computer Simulation of the Antiproton Channelling Process

In Lecture 4 we mentioned the first experimental studies of atomic collision processes involving antiprotons. Further development of the program of such studies relies significantly on calculations performed by the method of computer simulation.

From general considerations one may expect the charge sign effect, dealt with in Sect. 4.3, to be especially strong in the case of transmission through matter of channelled protons or antiprotons. The factors acting differently on channelled protons or antiprotons can be divided into two groups: of "kinematical" and of "dynamical" nature. In the first case qualitative differences are intended between the trajectories of motion of a proton and an antiproton in the channel. Channelled protons tend to occupy positions as close as possible to the middle of the channel, i.e., where the density of electrons in the medium and, consequently, the ionization stopping of the passing particle is the smallest. Antiprotons, on the contrary, tend to be as close as possible to the chains and planes forming the channel and thus not only pass a large part of time in the region of enhanced electron density, but are also more readily subject to dechannelling, owing to their interaction with the atomic nuclei of the medium. As to the differences of a dynamic character, they are ultimately understood to reduce to the difference between the elementary interactions between the proton or the antiproton and atoms of the medium; for instance, at one and the same velocity and given one and the same impact parameter a proton and an antiproton lose different fractions of their energy in such an elementary act.

Bearing in mind all these differences, we shall follow the motion of the proton and of the antiproton in the channel. For definiteness we consider the case of axial channelling; the z axis of the reference system will be directed along the channel axis. A particle trajectory is found from the equation of classical mechanics

$$m \frac{\mathrm{d}^2 r}{\mathrm{d}t^2} = F_{\mathrm{defl}} + F_{\mathrm{stop}} , \qquad (6.13)$$

where m is the particle mass, F_{defl} and F_{stop} are the deflecting force, with which the atoms forming the channel act on the particle, and the force of ionization stopping, respectively:

$$F_{\text{defl}}(r) = -\text{grad}\left[\sum_i u_i(r)\right],$$

$$F_{\text{stop}}(r) = -\frac{v}{v}\left(-\frac{dE}{dz}\right)_r. \tag{6.14}$$

Usually, the potential of the interaction of the particle transmitted with atoms of the walls of the channel is approximated by a sum of continuous potentials (Lindhard potentials) corresponding to a small number of the nearest rows of atoms. In this approximation, the longitudinal component of the deflecting force approaches zero, while its transverse component is independent of the particle coordinate z:

$$F_{\text{defl}}(r) = \{F_x(x, y), F_y(x, y), 0\}. \tag{6.15}$$

As to the force of ionization stopping (here, instead of the standard notation $-dE/dx$, we shall write it as $-dE/dz$), owing to the small curvature of the trajectory of the channelled particle, the transverse component of this force is negligible in comparison with its longitudinal component:

$$F_{\text{stop}}(r) \approx \left\{0, 0, -\left(-\frac{dE}{dz}\right)\right\}. \tag{6.16}$$

Variation of the stopping force along the particle trajectory $r = r(t)$ is determined by the profile of the electron density distribution $\rho_e(r)$ along the channel. A number of methods are known (for example, X-ray analysis of matter), which are used for determining this distribution. With quite a good precision one may consider that up to a certain radius R (approximately, in the range of the last occupied atomic shells) the electron density in the atoms of a crystal is the same as in the corresponding free atom; all the remaining space in between the spheres of radius R is filled with an homogeneous electron gas of constant density.

In modelling the passage of a particle – a proton or an antiproton – along the channel we shall consider that for a given impact parameter the particle's interaction with electrons of the inner atomic shells and with external electrons yields additive contributions to the energy losses of the particle:

$$\Delta E(b) = \Delta E_{\text{inner}}(b) + \Delta E_{\text{extern}}(b) \tag{6.17}$$

(although, naturally, only by convention can the atomic shells be divided into inner and external shells). For calculating the first term in (6.17) it is convenient to take advantage of the semiclassical approach, in which the motion of a passing particle is interpreted classically, while everything concerning the atoms of the medium is dealt with within the framework of quantum mechanics. Then, the energy loss of a particle due to interaction with each ith inner atomic shell is determined by the expression

$$\Delta E_i(b) = \sum_f (\varepsilon_f - \varepsilon_i) P_{i \to f}(b). \tag{6.18}$$

Here ε_i and ε_f are the energies of the initial and final states of the atomic electron (the final state may be not only an excited discrete level, but in the continuous spectrum, also), and $P_{i \to f}(b) = |A_{i-f}(b)|^2$ is the probability of the $i \to f$ transition induced by the particle passing with the impact parameter b. In the semiclassical approximation this probability amplitude is the integral over the time interval during which the particle is in the vicinity of the atom:

$$A_{i-f}(b) = \frac{1}{\hbar} \int_0^\infty e^{i\omega_{fi}t} \left\langle f \left| \frac{Z_1 e^2}{|r(t) - r_i|} \right| i \right\rangle dt, \tag{6.19}$$

where $\omega_{fi} = \frac{1}{\hbar}(\varepsilon_f - \varepsilon_i)$ is the transition frequency. In our case $Z_1 = \pm 1$ for the proton and the antiproton, respectively. The total energy loss of the particle due to excitation of the ionization of all-the inner shells is the sum of expressions (6.18):

$$\Delta E_{\text{form}}(b) = \sum_t \Delta E_t(b). \tag{6.20}$$

In the simplest version of the approach, the entire part of the trajectory $r(t)$ in the vicinity of the atom can be considered a straight line. Then, the energy loss due to interaction with the inner atomic shells is the same for the proton and the antiproton. If one also takes into account the bending of the trajectory of motion and/or the polarization of atomic shells by the incident particle, then precisely the so-called "dynamic" effect of the charge sign will arise, which was mentioned above.

For calculation of the second term in (6.17) one can take advantage of formulae from the dielectric theory of ionization stopping in the approximation of local density (see Sect. 3.4).

Thus, we have discussed nearly everything requiring definition for computing the trajectory of a channelled proton or antiproton. It remains, now, to apply the Monte Carlo method and determine the initial coordinates of the particle, $x(t = 0) = x_0$ and $y(t = 0) = y_0$, in the transverse plane, and to "follow" the particle along the channel by solving (6.13) numerically with these initial conditions. Among the various trajectories to be found by this method there may happen to be certain dangerous trajectories passing especially close to atoms of the lattice. The method of statistical simulation permits automatic tracing of such events and, as soon as the particle happens to be close to some atom at a distance smaller than a certain given critical distance r_c (usually it is chosen to be of the order of 0.01 Å), a change in the computation mode of the further motion of such a particle, if we assume a dechannelling act to have occurred.

Figure 6.16 presents an example of calculations performed by the method of computer simulation for channelled protons and antiprotons traversing a

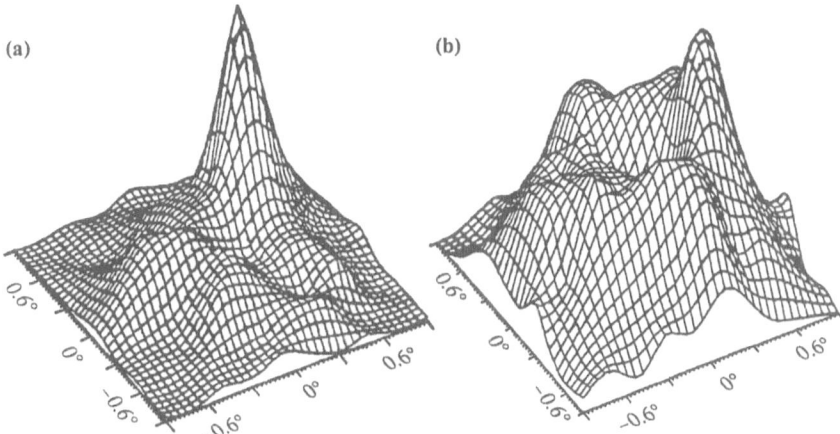

Fig. 6.16a, b. Simulation of the transmission of protons (a) and antiprotons (b) with an initial energy of 200 keV through a 3000 Å thick silicon foil along the ⟨100⟩ channel: angular distribution of particles at the exit of the channel [6.7]

thin silicon film. If the initial particle coordinates (x_0, y_0) are distributed uniformly at the entrance to the channel, their angular distributions at the exit from the channel differ qualitatively from each other; this is due to protons being concentrated as far as possible from the rows of atoms of the crystal in the case of channelling, while antiprotons, on the contrary, concentrate in the immediate vicinity of these rows. Computer experiments reveal that the kinematic and dynamic differences between the interactions of channelled protons and antiprotons with matter are also manifest in the energy distributions of particles transmitted through the film. On the whole, channelled antiprotons lose more energy than channelled protons, and this is primarily related to the difference between the trajectories of motion inside the channel (the "kinematic effect"). The difference between these losses, however, is also quite sensitive to the "dynamic" effects of the charge sign of the particle.

Part II

Interaction of Electromagnetic Radiation
with Matter

Seventh Lecture

7.1 Passage of γ Radiation Through Matter

General issues of the passage through matter of electromagnetic radiation (visible light, infrared, ultraviolet radiation, X-rays) are usually considered to be associated with optics and atomic physics. Here we examine the passage through matter of hard electromagnetic radiation with some photons having energies up to several tens and hundreds of MeV. In the following sections we shall consider two special issues from the field of modern studies in nuclear and elementary particle physics bordering on the problem of the transmission of electromagnetic radiation through matter.

Within a broad range of energies, the absorption cross section of γ quanta by the atoms of matter adds up of the cross sections of three processes: the photoelectric effect, the Compton effect, and the production of electron-positron pairs:

$$\sigma_{abs} = \sigma_{phot} + \sigma_C + \sigma_{pair} . \tag{7.1}$$

In the photoelectric effect all the energy of the γ quantum, $E_\gamma = h\nu$, is transferred to the atomic electron,

$$\gamma + A \rightarrow A^+ + e; \tag{7.2}$$

the kinetic energy of the outgoing electron, T_e, is higher, the lower the ionization potential I_i of the corresponding atomic shell is. The following simple relationship holds valid with a precision up to the recoil energy of the produced ion A^+, which even in the case of the lightest atoms is very small:

$$T_e = E_\gamma - I_i . \tag{7.3}$$

If one traces how the picture of the atomic photoelectric effect develops as the photon energy increases, two of its characteristic features arrest one's attention (Fig. 7.1): (a) the presence of sharp *absorption edges* at the ionization threshold of each subsequent shell or subshell; (b) the very rapid decrease of the cross section observed immediately after each value I_i as the energy E_γ increases. The combination of these two peculiarities gives rise to the characteristic serrated shape of the $\sigma_{phot}(E_\gamma)$ curve.

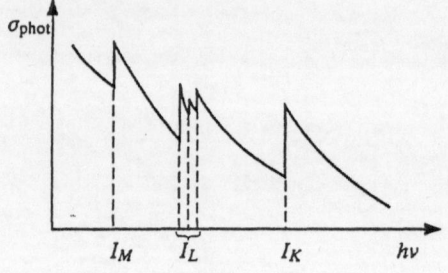

Fig. 7.1. Energy dependence of the cross section of the atomic photoelectric effect. The splitting of the absorption edge at the ionization threshold of the L shell

In quantum electrodynamics the atomic photoelectric effect is represented by Feynman diagrams, the simplest of which has the form:

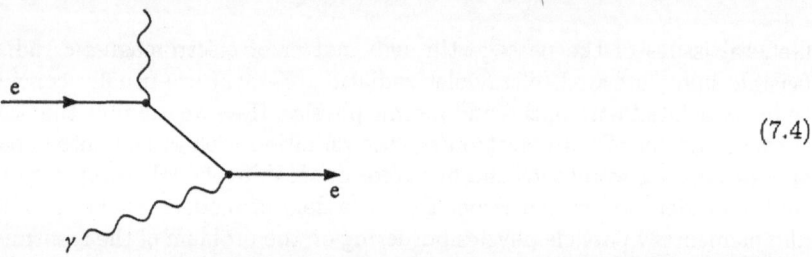

$$(7.4)$$

Here, the upper wavy line stands for the interaction of one of the atomic electrons with the total electric fields of the nucleus and the remaining electrons, and the lower vertex represents the actual process of adsorption of a photon by this electron. The same process of photon absorption by a free electron, which could be represented by the diagram

$$(7.5)$$

is totally forbidden, since it is impossible to simultaneously satisfy both the energy and momentum conservation laws for the free electron in process (7.5). Thus, the presence of a "third body", with which our electron is somehow connected, is a necessary condition for it to absorb a photon. The part of such a "third body" in the real situation (7.2) is played by the ion A^+, while the ionization potential I_i serves as a measure of its coupling to the electron. To judge how essential this coupling is for the photoelectric effect to be possible, one must compare I_i with the photon energy E_γ. When E_γ rises, the "boundedness" of the electron in the atom weakens, and when $E_\gamma \gg I_i$ it vanishes completely; the photoelectric effect is practically impossible in such conditions. Hence one can see that at one and the same photon energy the electrons of the inner shells (if E_γ is higher than the respective ionization thresholds, naturally) yield a significantly greater contribution to the photoabsorption process, than the electrons of the outer shells. It is also clear that at high E_γ the probability of the photoelectric effect should grow with the atomic number of the medium.

Theoretical calculations within the framework of quantum electrodynamics confirm these qualitative arguments. The cross section for the photelectric effect does actually fall rapidly as the photon energy increases:

$$\sigma_{\text{phot}} \sim \frac{Z^5}{E_\gamma^{7/2}}, \quad E_\gamma > I_i; \tag{7.6}$$

$$\sigma_{\text{phot}} \sim \frac{Z^5}{E_\gamma}, \quad E_\gamma \gg I_i; \tag{7.7}$$

and it rises sharply, as we see from these formulae, when transition occurs from light to heavy elements.

The dependences on E_γ and on Z of the cross section of the Compton effect is quite different. Here, the photon does not vanish when interacting with an electron of the medium, but in transferring only part of its energy to the electron undergoes scattering with enhancement of its wavelength. When $E_\gamma \gg I_i$, this process is described with the aid of the Feynman diagrams for a free electron:

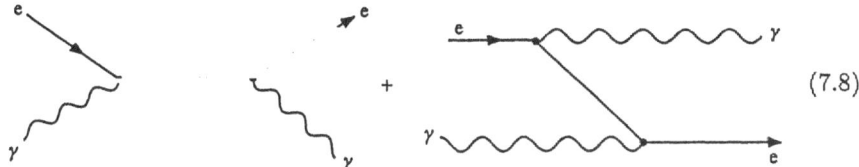

$$\tag{7.8}$$

Here, the total cross section for the Compton scattering of γ quanta on an atom is proportional to the number of electrons in the atom, i.e. to the atomic number of the substance, Z. Quantum electrodynamics permits us to obtain an analytic, although quite cumbersome, expression for the cross section of process (7.8); in the limit of $E_\gamma \gg m_e c^2$ this cross section is inversely proportional to the energy of the γ quantum:

$$\sigma_{\text{C}}\Big|_{E_\gamma \gg m_e c^2} \sim \frac{Z}{E_\gamma}. \tag{7.9}$$

When $E_\gamma > 2m_e c^2$, transformation of the γ quantum into an electron-positron pair becomes possible from the point of view of energy conservation. But, like the absorption of a photon by an electron, this process cannot take place without the additional body. The role of such a body may be assumed by an electron or by the atomic nucleus. Owing to the electron mass being small, the pair-production threshold, as compared with the sum of the electron and positron masses, rises sharply in the first case:

$$E_\gamma\Big|_{\gamma+e\to e+e+e^+} > 4m_ec^2 \;; \tag{7.10}$$

the cross section of pair production on the atomic electrons, reduced to one atom, is proportional to Z. Pair production on the atomic nucleus is more substantial. In the case of a "bare" nucleus, the corresponding cross section is proportional to Z^2. Actually, screening of the electric field of the nucleus by the atomic electrons somewhat weakens this dependence on Z and also affects the character of the energy dependence of the cross section. In the energy interval $m_ec^2 \ll E_\gamma < m_ec^2/(\alpha Z^{1/3})$ the pair-production cross section follows the formula

$$\sigma_{\text{pair}} \sim Z^2 \ln \frac{E_\gamma}{m_ec^2}\,, \tag{7.11}$$

and then its rise with the energy slows down.

Besides the absorption cross section, the total cross section of interaction between γ quanta and an atom includes the integral cross section σ_{el} of elastic scattering, which is also termed coherent or Rayleigh scattering. Unlike Compton scattering, which is accompanied by ionization or excitation of the atom, coherent photon scattering takes place on the atom as a whole and leaves its internal state unchanged. The contribution of coherent scattering to the total cross section is not significant within the entire range of energies E_γ, it is most noticeable in the regions where the contributions given by the photoelectric and Compton effects overlap (Fig. 7.2).

The main components of the cross section for absorption of γ quanta on the atoms of a substance, (7.1), behave differently, depending on Z and on the energy E_γ. Therefore, the $\sigma_{\text{abs}}(E_\gamma)$ curves for light and heavy substances

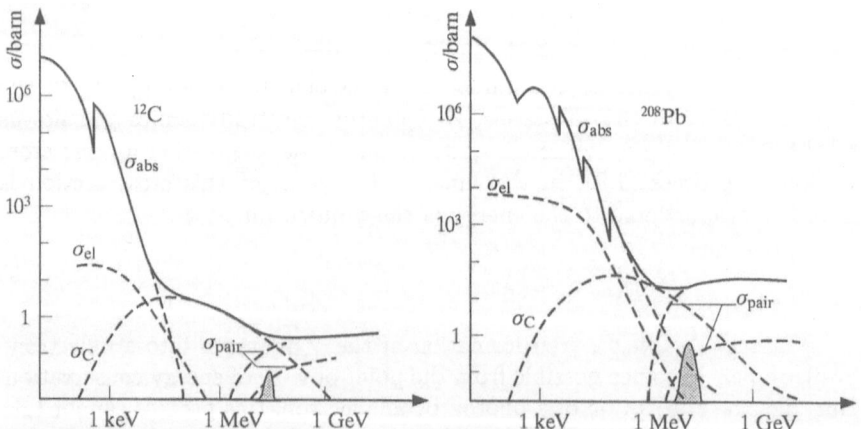

Fig. 7.2. Contributions of different mechanisms to the total cross section for interaction of photons with carbon and lead atoms [7.1]. The *shaded area* demonstrates the total cross section of absorption of γ quanta on the atomic nucleus of the atom

differ in shape significantly. In the case of large Z, the region in which the Compton effect is essential is narrow, owing to the dominant contribution of the cross section of the photoelectric effect at $E_\gamma \leq m_e c^2$ and to the contribution of the pair production cross section at $E_\gamma \gg m_e c^2$, and the $\sigma_{abs}(E_\gamma)$ curve has a clear minimum at γ quantum energies of several MeV (note the logarithmic scale in Fig. 7.2). In substances with small Z the region where the Compton effect cross section provides a dominant contribution to the total cross section is broad and the logarithmic rise of the cross section $\sigma(E_\gamma)$ related to pair production starts at significantly higher energies.

The cross section $\sigma(E_\gamma)$ determines the photon absorption coefficient

$$\mu(E_\gamma) = n_0 \sigma(E_\gamma), \tag{7.12}$$

which characterizes the change of intensity in a photon beam of a given energy E_γ transmitted through a layer of matter:

$$I(x) = I_0 e^{-\mu(E_\gamma)x}. \tag{7.13}$$

7.2 Interaction of γ Quanta with the Atomic Nuclei of Matter. The Mössbauer Effect

Let us now compare the cross sections of the atomic processes considered above with the cross sections of nuclear reactions caused by γ quanta. The absorption cross section at the minimum of the $\sigma_{abs}(E_\gamma)$ curve is close, by an order of magnitude, to the cross section of the Compton effect at $E_\gamma \approx m_e c^2$, which in turn can be estimated simply from dimensionality arguments:

$$\sigma_C \sim Z \left(\frac{e^2}{m_e c^2} \right)^2. \tag{7.14}$$

The second factor is the square "classical electron radius" $r_e^2 = (2.8 \times 10^{-13}\text{ cm})^2$, which is a quantity of the order of magnitude of the geometric cross section of the nucleus, πR_{nucl}^2. The cross sections of nuclear electromagnetic processes occurring in various situations may also be much smaller or much larger than this value. Let us estimate the nuclear absorption cross section of γ quanta, $\sigma_{nucl}(E_\gamma)$, at the maximum of the giant resonance, i. e., at $E_\gamma = 15$–25 MeV. The integral photoabsorption cross section in the region of the giant resonance is known, from nuclear physics, to be determined by the formula

$$\int \sigma_{nucl}(E_\gamma)\, dE_\gamma \approx 60 \frac{NZ}{A}\text{ MeV mbarn}. \tag{7.15}$$

In the case of nuclei in the region of $A = 50$ this amounts to about 1 MeV barn. Assuming the width of the giant resonance to be 5–8 MeV we obtain for the cross section σ_{nucl} at the maximum of the giant resonance a value of the order of 0.1 barn $= 10^{-25}$ cm^2.

The absorption cross section of γ quanta on a nucleus may reach very much higher values at certain sharp peaks corresponding to resonance transitions of the nucleus from the ground state to an excited state. In this case one can take advantage of the resonance formula

$$\sigma_{\text{nucl}} = g \frac{\pi}{k_\gamma^2} \frac{\Gamma^2}{(E_\gamma - E_{\text{res}})^2 + \frac{1}{4}\Gamma^2}, \tag{7.16}$$

where g is a statistical factor of the order of unity, E_{res} is the center of gravity of the resonance line in the photoabsorption spectrum, Γ is the width of the excited state to which transition of the nucleus, due to absorption of a γ quantum, occurs. From (7.16) it is seen that the maximum value of the cross section σ_{nucl} is enormous (when $E_\gamma = E_{\text{res}}$) if the lowest levels of the nucleus are excited. For example, $(\sigma_{\text{nucl}})_{\text{max}} \sim 10^{-19}\,\text{cm}^2$ at $E_{\text{res}} = 0.1$ MeV. We must not, however, forget that the cross section of resonance photoabsorption is so large only within a very narrow – down to $10^{-8}\,\text{eV}$ – range of energies. The integral photoabsorption cross section corresponding to an individual resonance is negligible (here, $\sim 10^{-19}\,\text{cm}^2 \times 10^{-8}\,\text{eV} = 10^{-6}\,\text{MeV mbarn}$) and, consequently, when a sample is irradiated with a flux of γ quanta with a continuous spectrum, their undergoing resonance absorption on a nucleus plays no part in the attenuation of this flux.

It is different with the absorption of monochromatic γ quanta of an energy precisely equal to the resonance energy E_{res}, as is the case in optics: an excited atom of matter undergoing a transition $|n\rangle \rightarrow |0\rangle$ emits a photon of a certain energy $\hbar\omega = \varepsilon_n - \varepsilon_0$, while another atom absorbs this photon, thus undergoing the inverse transition $|0\rangle \rightarrow |n\rangle$. In nuclear physics, however, such a situation is impossible in normal conditions: when emitting or absorbing a γ quantum, a nucleus acquires such a significant recoil energy that the system is no longer in resonance. The recoil energy, $R = k_\gamma^2/(2M_A)$, turns out to be considerably larger than the actual width Γ of the resonance.

Now, consider the relation between these two quantities, making use of the example of the ^{57}Fe nucleus (Fig. 7.3). The width of the first excited state $3/2^-$, which we shall calculate from the lifetime of this state, $\tau = \hbar/\Gamma$, is a quantity of the order of 10^{-8} eV. The energy shift of the γ quantum emitted when this state undergoes de-excitation amounts to $R = (0.014\,\text{MeV})^2/2 \times 57 \times 938\,\text{MeV} \approx 2 \times 10^{-3}$ eV, i.e., it exceeds the width of the resonance by many orders of magnitude.

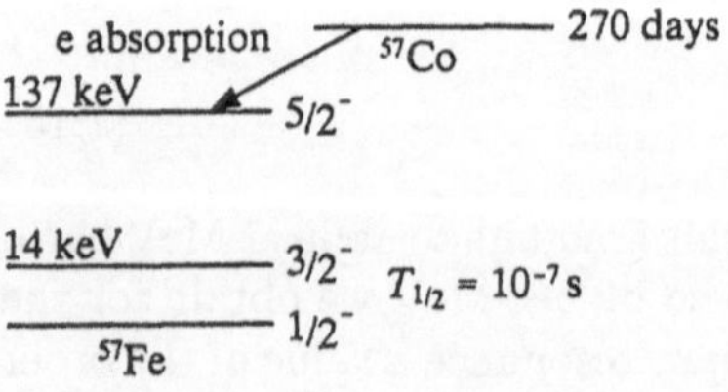

Fig. 7.3. Arrangement of the lowest levels of the ^{57}Fe nucleus

However, it is possible to observe resonance absorption of γ quanta by nuclei owing to the Mössbauer effect. This effect was discovered in 1958 during studies of nuclear resonance absorption in crystals under conditions of strong cooling both of the source and of the absorber. Courses of general physics provide a qualitative explanation of the Mössbauer effect: owing to the strong coupling between an individual atom of a crystal and the whole crystalline lattice, the recoil momentum $k_R = \pm k_\gamma$ arising when a γ quantum is either absorbed or emitted may be taken up, not by this separate atom, but by the crystal as a whole or at least by a very large (microscopically large) group of atoms. Now, for calculating the energy shift of the γ quantum in the formula $R = k_\gamma^2/2M$ it is necessary to substitute a large mass for the mass of a sole nucleus M_A. This just means that the emission or the absorption of a γ quantum occurs practically without any energy shift.

The issue of how often such an event takes place, i.e., the question concerning the relative probability of events without the energy of the γ quantum being taken up by the recoil, remains outside the scope of this explanation. To estimate both the value and the temperature dependence of this relative probability we must consistently put together our picture of γ quanta passing through the crystal and their absorption (emission) by the nuclei of atoms making up the crystal. According to Einstein's model each atom of a solid represents an independent harmonic oscillator, which has its own natural oscillation frequency ω, i.e., the energy $\hbar\omega$ of an oscillation quantum. The energy of the nth state (containing n quanta) is given by the formula $E_n = \hbar\omega(n+1/2)$, where the quantity $E_0 = 1/2\hbar\omega$ is the energy of vibrations in the ground state. The mean number of quanta, $\overline{n}$, of definite energy $\hbar\omega$ depends on the temperature of the solid and is determined by the Bose–Einstein formula:

$$\overline{n} = \frac{1}{\exp(\hbar\omega/kT) - 1}. \qquad (7.17)$$

Unlike in the case of Einstein's model, the oscillations of atoms in a crystal are interconnected, so there actually exists a whole quasicontinuous spectrum of natural oscillation frequencies of the lattice. The frequency distribution is characterized by the spectral density of oscillations, $D(\omega)$, the shape of which determines a whole series of various physical properties of the crystal starting with its heat capacity and going up to the temperature dependence of the Mössbauer effect. The shape of the spectral density $D(\omega)$ is not trivial (Fig. 7.4) and is peculiar to each substance, but in many cases it can be approximated with the aid of the Debye model

$$D(\omega)d\omega = \begin{cases} \text{const } \omega^2 d\omega, & \omega \leq \omega_D; \\ 0, & \omega > \omega_D. \end{cases} \qquad (7.18)$$

In this model the Debye frequency ω_D, i.e., the maximum oscillation frequency of the lattice, or the Debye temperature T_D, related to it by the relation

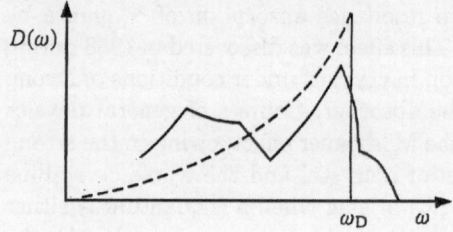

Fig. 7.4. Distribution of spectral density for lattice oscillations $D(\omega)$ and its approximation by the Debye model

$$\hbar\omega_D = kT_D, \tag{7.19}$$

serves as the structural characteristic of each crystalline substance. The values of ω_D and T_D for various substances can be found in reference books. Thus, for example, the Debye temperature for the lattice of metallic iron amounts to $467\,\mathrm{K}$ reduced to the maximum energy of the oscillation quantum, which gives $\hbar\omega_D \approx 5 \times 10^{-2}\,\mathrm{eV}$. In the Debye model, the mean number of oscillation quanta of the lattice for any frequency ω within the interval $0 < \omega < \omega_D$ is given by (7.17).

Let us first consider the Mössbauer effect from the standpoint of the simplest Einstein model. Consider a nucleus emitting a γ quantum and, thus, undergoing transition from one state of its internal motion, $|i\rangle$, to another state, $|f\rangle$. When this is a free nucleus, the energy of the emitted γ quantum is uniquely determined by the difference between the levels of the nucleus:

$$E_\gamma = \varepsilon_i - \varepsilon_f - R, \tag{7.20}$$

(where $R = k_\gamma^2/2M_A$ is the recoil energy), and the transition probability is just the matrix element of the electromagnetic interaction operator $\hat{H}_\gamma$ in between the two states of the internal motion of the nucleus:

$$\left(W_{fi}\right)_{\mathrm{free}} \sim \left|\langle f|\hat{H}_\gamma|i\rangle\right|^2. \tag{7.21}$$

On the other hand, if the nucleus under consideration is part of the lattice of the crystal, then the initial and final states of the whole system are no longer characterized only by indices of the internal state of the nucleus itself, but also by additional quantum numbers related to the motion of the nucleus as a whole inside the crystal: $|i\rangle \Leftrightarrow |i, n_i\rangle$; $|f\rangle \Leftrightarrow |f, n_f\rangle$. In Einstein's model such are the numbers of quanta of the oscillations of our nucleus, n_i and n_f, before and after emission of the γ quantum. The energy of the emitted γ quantum depends on how much these numbers change when the nuclear transition $|i\rangle \rightarrow |f\rangle$ occurs:

$$E_\gamma = \varepsilon_i - \varepsilon_f + \hbar\omega(n_i - n_f). \tag{7.22}$$

At the same time the transition probability is also calculated by taking into account the change in state of the nucleus in the lattice:

$$W(i, n_i \rightarrow f, n_f) = (W_{fi})_{\mathrm{free}}\left|\langle n_f|\mathrm{e}^{i\boldsymbol{k}_\gamma \cdot \boldsymbol{r}}|n_i\rangle\right|^2. \tag{7.23}$$

The second factor in this expression plays the part of an inelastic form factor [compare it, for instance, with (2.16)], its value depends on the momentum of the γ quantum, k_γ. We have considered nuclear transition in the crystal involving the emission of a γ quantum; in essence, all the above is also valid, without alteration, for the absorption of a γ quantum.

Thus, in considering nuclear transition within the framework of the Einstein model we deal with three energy parameters: the difference between levels, $\varepsilon_i - \varepsilon_f$ (in the example of ^{57}Fe presented above this amounts to 14 keV), the width of the level, Γ ($\sim 10^{-8}$ eV, in the same example), and the energy of a quantum of atomic oscillations in the lattice, $\hbar\omega$ (we shall tentatively set $\hbar\omega \approx 5 \times 10^{-2}$ eV, which corresponds to the Debye frequency in metallic iron). In normal conditions, the energy resolution ΔE_γ of the detectors used in detecting emitted γ quanta is significantly worse than required for resolving the structure of a line related to various possible changes in the number of oscillation quanta of an atom in the lattice undergoing the transition $|i, n_i\rangle \rightarrow |f, n_f\rangle : \Delta E_\gamma \gg \hbar\omega$. In the case of such detection conditions the total emission probability of a γ quantum in the nuclear transition $|i\rangle \rightarrow |f\rangle$ reduced to a definite intial state of the system, $|i, n_i\rangle$, should be computed as the sum of the transition probabilities over all possible final states of the atom in the lattice:

$$W_{fi}\Big|_{\Delta E_\gamma \gg \hbar\omega} = (W_{fi})_{\text{tot}} = \sum_{n_f} W(i, n_i \rightarrow f, n_f). \tag{7.24}$$

Substituting, here, (7.23) and taking into account the quantum-mechanical identity

$$\sum_{n_f} \left| \langle n_f | e^{i k \cdot r} | n_i \rangle \right|^2 = 1, \tag{7.25}$$

we obtain, as was to be expected, that the total probability of a spontaneous γ transition occuring in a nucleus is the same whether the nucleus is located inside a crystal or whether it is free: $(W_{fi})_{\text{tot}} = (W_{fi})_{\text{free}}$.

The situation is quite different, when the emitted γ quantum is detected by the method of resonance absorption, i. e., under conditions of the Mössbauer effect. The sensitivity of such a "device" is so high that even in the case of a minimum shift in the γ quantum energy ($E_\gamma = \varepsilon_i - \varepsilon_f \pm \hbar\omega$) the system stops being in resonance. The energy of the emitted γ quantum, E_γ, is exactly equal to the difference between the levels, $\varepsilon_i - \varepsilon_f$, only in the case of "diagonal" transitions, when the number of oscillation quanta remains unaltered: $n_f = n_i$. The probability of such a transition is always smaller than the probability for a γ transition to occur in a free nucleus:

$$W(i, n \rightarrow f, n) = (W_{fi})_{\text{free}} \left| \langle n | e^{i k_\gamma \cdot r} | n \rangle \right|^2 < (W_{fi})_{\text{free}}. \tag{7.26}$$

A most important characteristic applied in the theory of the Mössbauer effect is the emittance of a fraction of γ quanta without energy being transferred to the lattice (subsequently, we shall simply say "emitted without recoil", implying that it is the energy transfer that is negligible, not the momentum transferred to the crystal).

We shall call this fraction of the number of γ quanta, or equivalently the relative probability of recoilless emission of a γ quantum,

$$f = \frac{(W_{fi})_{\text{recoilless}}}{(W_{fi})_{\text{free}}}, \tag{7.27}$$

the *attenuation factor*. When it is calculated with the thermal oscillations of the lattice taken into account within the framework of the Debye model, it is also known as the Debye–Waller factor $f_{\text{D-W}}$. Our ultimate goal in this section consists precisely in expressing the Debye–Waller factor in terms of parameters of the lattice and in revealing its dependence on the temperature of the crystal.

When a nucleus emitting a γ quantum is in a certain stationary state of its oscillatory motion, the attenuation factor f may be factorized out of equation (7.26):

$$f_n = \left| \langle n | e^{i k_\gamma \cdot r} | n \rangle \right|^2. \tag{7.28}$$

First, we shall calculate the factor f_n for the case of ground-state oscillations of a nucleus inside the lattice. Directing the x axis along the trajectory of the outgoing γ quantum we have for the corresponding matrix element

$$\langle 0 | e^{i k_\gamma \cdot r} | 0 \rangle = \langle 0 | e^{i k_\gamma \cdot r} | 0 \rangle$$

$$= \int_{-\infty}^{\infty} \psi_{n=0}^*(x) e^{i k_\gamma x} \psi_{n=0}(x) \, dx. \tag{7.29}$$

The explicit form of the wave functions $\psi_n(x)$ describing the stationary states of a harmonic oscillator is well known: $\psi_{n=0}(x) = 1/(\pi^{1/4} x_0^{1/2}) \times \exp[-1/2(x/x_0)^2]$, where the oscillatory parameter x_0 is determined by the natural frequency and mass of the oscillator: $x_0 = \sqrt{\hbar/M\omega}$. The parameter x_0 defines the amplitude scale of oscillations. Thus, the mean square amplitude of oscillations in the stationary state $\psi_n(x)$ is calculated by the formula:

$$\langle x^2 \rangle_n = \left(n + \frac{1}{2} \right) x_0^2 = \left(n + \frac{1}{2} \right) \frac{\hbar}{M\omega}. \tag{7.30}$$

Substituting (7.29) into (7.28) and integrating directly over x we obtain

$$\langle 0 | e^{i k_\gamma \cdot r} | 0 \rangle = \exp\left(-\tfrac{1}{4} k_\gamma^2 x_0^2 \right) = \exp\left(-\tfrac{1}{2} k_\gamma^2 \langle x^2 \rangle_{n=0} \right). \tag{7.31}$$

With the aid of the expression for the oscillatory wave functions $\psi_n(x)$, taken from quantum mechanics, one can readily verify that formula (7.31) is valid not only for the lowest but also for any other state of the oscillator:

$$\langle n | e^{i k_\gamma \cdot r} | n \rangle = \exp\left(-k_\gamma^2 \langle x^2 \rangle_n \right). \tag{7.32}$$

Hence we find that the attenuation factor f_n for the $|i\rangle \rightarrow |f\rangle$ transition probability is given by the expression

$$f_n = \exp\left(-k_\gamma^2 \langle x^2 \rangle_n\right). \tag{7.33}$$

The result obtained concerns an individual harmonic oscillator in a certain stationary state. Let us generalize it to an ensemble of oscillators in a certain given (generally mixed) state by expressing the mean value of the attenuation factor f via the mean square amplitude of nuclear oscillations in the given state:

$$f = \exp\left(-k_\gamma^2 \langle x^2 \rangle\right). \tag{7.34}$$

It is now an opportune moment to digress from consistent calculations and to examine relation (7.34) from the point of view of the qualitative conditions for the Mössbauer effect to exist. We see that the probability for a nucleus to undergo recoilless emission of a γ quantum depends on the relationship between the radiation wavelength $\lambda = 1/k_\gamma$ and the average dimensions of the region within which the motion of the nucleus is localized. A necessary condition for the Mössbauer effect to exist is that these average dimensions should be small compared with the wavelength:

$$\langle x^2 \rangle \ll \lambda^2. \tag{7.35}$$

Condition (7.35) is not fulfilled in gases and liquids, since a nucleus emitting a γ quantum is in no way localized in space. The issue of the extent to which condition (7.35) is satisfied when applied in the case of crystals requires a more detailed consideration. From (7.34) it is seen that the probability of recoilless emission of γ quanta falls as the temperature rises, since in this case the amplitude of nuclear oscillations increases. On the other hand, even at the lowest temperatures, in the vicinity of absolute zero, the attenutation factor $\exp(-k_\gamma^2 \langle x^2 \rangle)$ does not reach unity, owing to the existence of ground-state oscillations of the nuclei in the lattice.

We shall now return to numerical computations within the Einstein model and calculate the mean square oscillation amplitude $\langle x^2 \rangle$ occurring in (7.34), for an ensemble of oscillators in thermal equilibrium with the surrounding medium. To this end, it is sufficient, as one can see from (7.30), merely to know the mean value of the number of oscillation quanta, $\bar{n}(T)$, at the given temperature:

$$\langle x^2 \rangle_T = \left[\bar{n}(T) + \frac{1}{2}\right] \frac{\hbar}{M\omega}. \tag{7.36}$$

In turn, the mean value $\bar{n}(T)$ is given by the Bose–Einstein formula (7.17). As a result we obtain

$$\langle x^2 \rangle_T = \frac{\hbar}{M\omega}\left(\frac{1}{e^{\hbar\omega/kT}-1} + \frac{1}{2}\right). \tag{7.37}$$

We have now exhausted the possibilities of the simple Einstein model, in which each atom represents an independent harmonic oscillator of definite frequency ω.

To take into account the weights of different natural frequencies in the case of bound oscillations of atoms of the crystalline lattice we turn to the Debye model and average (7.37) over the frequency spectrum (7.18):

$$\langle x^2 \rangle_T \implies \frac{\int\limits_0^{\omega_D} \frac{\hbar}{M\omega}\left(\frac{1}{\exp(\hbar\omega/kT)-1} + \frac{1}{2}\right)\omega^2 d\omega}{\int\limits_0^{\omega_D} \omega^2 d\omega}. \tag{7.38}$$

Upon performing simple calculations and substituting the result obtained into (7.34) we arrive at the final expression for the attenuation factor, the Debye–Waller factor:

$$f_{D\text{-}W}(T) = \exp\left\{ -\frac{3}{2}\frac{R}{\hbar\omega_D}\left[1 + 4\left(\frac{T}{T_D}\right)^2 \int\limits_0^{T_D/T} \frac{u\,du}{e^u - 2}\right]\right\}, \tag{7.39}$$

where $R = \hbar^2 k_\gamma^2/2M$ is the recoil energy due to the emission of a γ quantum by the free nucleus.

Expression (7.39) contains an indefinite integral, which can be calculated analytically only in the limit cases in which the temperature of the crystal is much lower or much higher than the Debye temperature:

$$I\left(\frac{T_D}{T}\right) = \int\limits_0^{T_D/T} \frac{u\,du}{e^u - 1} = \begin{cases} \frac{\pi^2}{6}, & T \ll T_D; \\ \frac{T_D}{T}, & T \gg T_D. \end{cases} \tag{7.40}$$

Accordingly, in these extreme conditions we obtain for the Debye–Waller factor

$$f_{D\text{-}W}(T) = \begin{cases} \exp\left(-\frac{3}{2}\frac{R}{\hbar\omega_D}\right), & \text{if } T \ll T_D; \\ \exp\left(-6\frac{R}{\hbar\omega_D}\frac{T}{T_D}\right), & \text{if } T \gg T_D. \end{cases} \tag{7.41}$$

Hence it is seen that for reliable observation of the Mössbauer effect it is necessary:

(a) to select a γ-emitter in the region of heavy nuclei with low transition energies (small R);
(b) to choose crystals with high Debye temperatures T_D;
(c) to cool the emitter and absorber down to as low a temperature as possible.

However, (as pointed out at the very beginning of our discussion) the presence of zero oscillations of the lattice prevents 100 % recoilless emission or absorption of γ quanta even at absolute zero.

To conclude, we shall estimate the Debye–Waller factor for, for example, the $3/2 \rightleftharpoons 1/2$ transition in the ^{57}Fe nucleus in the lattice of metallic iron. At low temperatures, $T \ll T_D = 467\,\text{K}$, we have

$$f_{\text{D-W}}\Big|_{T \ll T_D} = \exp\left(-\frac{3}{2}\frac{2 \times 10^{-3}\,\text{eV}}{5 \times 10^{-2}\,\text{eV}}\right) \approx 0.95. \tag{7.42}$$

This is a very high value. Numerical calculation with (7.39) reveals that it remains sufficiently high even at room temperatures: $f_{\text{D-W}}(300\,\text{K}) \approx 0.75$. Therefore the Mössbauer effect can be readily observed in iron up to very high temperatures.

Eighth Lecture

8.1 Rotation of the Plane of Polarization of Light as an Effect Due to Parity Violation in Atoms

The discovery of intermediate bosons, which was one of the most outstanding achievements of physics in the 1980s, was actually preceded by long-standing theoretical and experimental studies in the field of weak interactions and, in particular, by the investigation of neutral weak currents. The existence of such currents in nature was very desirable from the the point of view of the development of elementary particle theory; however, for a long time they could not be found. All weak interaction processes known by the beginning of the 1970s only confirmed the existence of charged weak currents: in all such processes either the charge state of a lepton is changed, like, for example, in μ capture,

$$(8.1)$$

or the lepton pair produced carries a nonzero total charge, like, for instance, in the case of the β decay of the neutron:

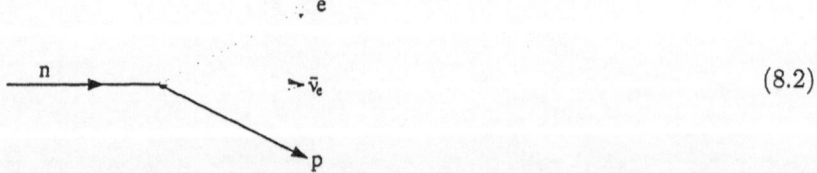

$$(8.2)$$

In modern diagram language both processes (8.1) and (8.2) are interpreted as the exchange of a charged intermediate boson; for example,

(8.3)

Processes of neutrino scattering from hadrons could serve as a manifestation of neutral currents. That is precisely how they were first revealed (CERN, 1973), by the absence of muons in a significant part of events corresponding to inelastic interactions of muon neutrinos and antineutrinos with nuclei:

$$\nu_\mu + A \rightarrow \nu_\mu + \text{hadrons},$$
$$\bar{\nu}_\mu + A \rightarrow \bar{\nu}_\mu + \text{hadrons}.$$

(8.4)

The existence of a neutral intermediate boson permits the interpretation of such processes within the framework of the exchange mechanism, which is universal for all sorts of interactions; for example,

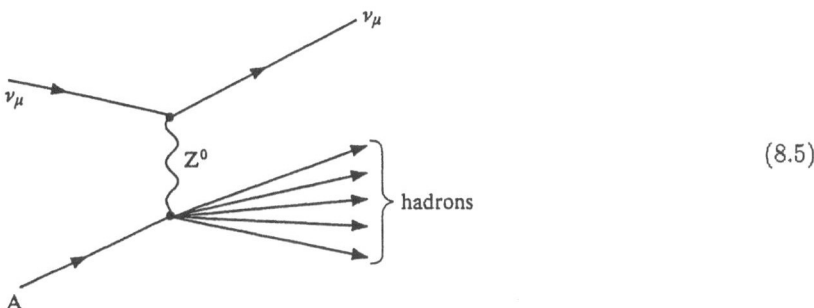

(8.5)

Another manifestation of neutral currents consists in the effects of parity violation in atomic systems. Let us consider the interaction between an electron and a proton. According to quantum electrodynamics it is due to the exchange of a photon:

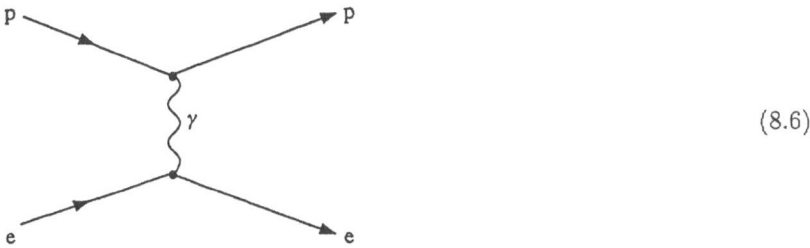

(8.6)

The presence of neutral weak currents (the existence of a neutral intermediate boson) requires adding to this diagram another one:

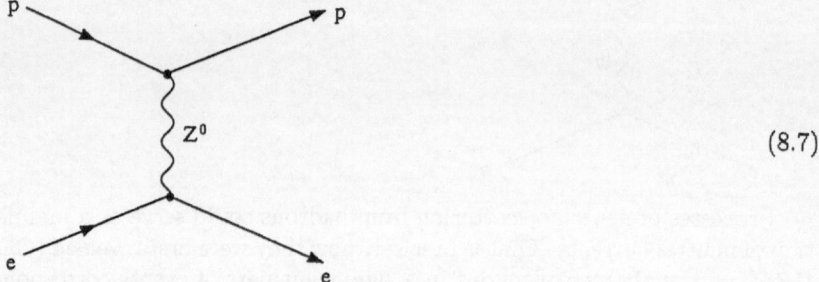

$$(8.7)$$

Thus, the total Hamiltonian of electron–proton interaction is the sum of two terms:

$$\hat{H}_{\text{ep}} = \hat{H}_0 + \Delta\hat{H}. \tag{8.8}$$

Although the second summand $\Delta\hat{H}$ describing the weak interaction (8.7) is negligible in comparison with the main summand $\hat{H}_0$ corresponding to the electromagnetic interaction, it nevertheless introduces into the Hamiltonian $\hat{H}_{\text{ep}}$ an absolutely new property: owing to this summand parity is not conserved in the interaction of charges.

Imagine an atom with a single valence electron, for example, an atom of an alkali metal (Fig. 8.1). In atomic physics all the stationary states of an atom (with the exception of the states of a hydrogen atom representing a case of degeneracy) are states of certain parities. Such, also, are the states of our atom, $ns_{1/2}$ and $np_{1/2}$. If we take into account the interaction $\Delta\hat{H}$ violating parity, these states (for greater generality we shall also denote them by $|+\rangle$ and $|-\rangle$) may only serve as a basis for finding the true wave functions of the two states considered:

$$
\begin{aligned}
|a\rangle &= \alpha|+\rangle + \beta|-\rangle, \\
|b\rangle &= \beta|+\rangle + \alpha|-\rangle.
\end{aligned}
\tag{8.9}
$$

Here, the coefficient α practically coincides with unity, whereas the coefficient β characterizing the admixture of states of opposite parity is very small; perturbation theory estimates,

$$\beta \approx \frac{\langle +|\Delta\hat{H}|-\rangle}{\varepsilon_0(-) - \varepsilon_0(+)}, \tag{8.10}$$

show that under usual conditions the coefficient β is of the order of 10^{-16}.

$\varepsilon_0(-)$ ——— $np_{1/2} \equiv |-\rangle$ ——— $|b\rangle$

$\varepsilon_0(+)$ $\downarrow E1$ $\downarrow E1 + M1$

——— $ns_{1/2} \equiv |+\rangle$ ——— $|a\rangle$

(a)   **(b)**

Fig. 8.1. Diagram of electromagnetic transition between the lowest levels of the valence electron without (a) and with (b) parity conservation being taken into account

Since parity is violated, selection rules based on parity must be excluded from the set of selection rules regulating electromagnetic transitions between atomic states. In the case under consideration, this means that not only is the $E1$ transition possible between the atomic states shown in Fig. 8.1, but, strictly speaking, the $M1$ transition can also take place. Let $\hat{H}_\gamma$ be the Hamiltonian for interaction of an atom with the electromagnetic radiation responsible, for example, for the processes of photon emission and absorption. The probability for the atom to undergo spontaneous electromagnetic transition from state to state is determined by the matrix element of the operator

$$\langle b|\hat{H}_\gamma|a\rangle = \langle E1\rangle_{ba} + \beta\langle M1\rangle_{ba}\,, \tag{8.11}$$

which we have divided into two terms: the first corresponds to the usual selection rules, the second, which is proportional to a small parameter β, is forbidden in the usual approach and has arisen here as the effect of parity nonconservation. Naturally, it is absolutely impossible to feel such a small addition by simply measuring the electromagnetic transition probability $W_{b\to a} \sim |\langle b|\hat{H}_\gamma|a\rangle^2$.

We recall, however, that one of the characteristic properties of electromagnetic radiation is its polarization: light can be linearly polarized, circularly polarized, etc. Turning to quantum electrodynamics, it is possible to show that the parity-allowed and parity-forbidden summands of the matrix element (8.11) combine in transitions to result in right- and left-polarized light being emitted differently:

$$\langle b|\hat{H}_\gamma^{\text{right}}|a\rangle = \langle E1\rangle_{ba} \pm \beta\langle M1\rangle_{ba}\,,$$
$$\langle b|\hat{H}_\gamma^{\text{left}}|a\rangle = \langle E1\rangle_{ba} \mp \beta\langle M1\rangle_{ba}\,. \tag{8.12}$$

Thus, parity violation may manifest itself as the difference between the probabilities of spontaneous electromagnetic transitions leading to the emission of photons of opposite circular polarizations. The resultant effect consists in spontaneously emitted radiation exhibiting a very small (but actually non-zero) circular polarization:

$$P = \frac{\left|\langle b|\hat{H}_\gamma^{\text{right}}|a\rangle\right|^2 - \left|\langle b|\hat{H}_\gamma^{\text{left}}|a\rangle\right|^2}{\left|\langle b\hat{H}_\gamma^{\text{right}}|a\rangle\right|^2 + \left|\langle b|\hat{H}_\gamma^{\text{left}}|a\rangle\right|^2} \approx \beta\frac{2\text{Re}\{|\langle E1\rangle_{ba}^*\langle M1\rangle_{ba}|\}}{|\langle E1\rangle_{ba}|^2}\,. \tag{8.13}$$

As we see, this circular polarization is proportional to the coefficient β characterizing the admixture of the state of opposite parity.

The probabilities of induced electromagnetic transitions (say, with absorption of a photon) are proportional to the probabilities of spontaneous transitions between the respective levels. From the above consideration it follows that owing to parity violation the excitation probabilities of individual atomic levels due to the atom absorbing right and left circularly polarized light, strictly speaking, do not coincide with each other, $W_{a\to b}^{(\text{right})} \neq W_{a\to b}^{(\text{left})}$; naturally, the same also holds valid for the respective oscillator forces,

$f_{a \to b}^{(\text{right})} \neq f_{a \to b}^{(\text{left})}$. Hence follows an essentially new situation concerning the transition of light through matter: taking into account parity violation in atoms, the refractive indices of a medium for right- and left-polarized light do not coincide with each other: $n^{(\text{right})} \neq n^{(\text{left})}$. From the formula

$$n(\omega) = 1 + \frac{2\pi e^2 N}{m_e} \sum_n \frac{f_n}{\omega_0^2 - \omega^2 + \mathrm{i}\omega\gamma_n} \qquad (8.14)$$

for the refractive index of a gas it can be seen that the effect should be particularly noticeable at frequencies close to those natural frequencies of atoms, which correspond to the excitation of states with large admixtures of the parity-forbidden component.

We have, here, conditions leading to the phenomenon, well-known in optics, of optical activity of a medium. When linearly polarized light passes through such a medium, its polarization plane is rotated. The angle of rotation is given by the Fresnel formula

$$\psi = \pi \frac{l}{\lambda} \mathrm{Re}\{n^{\text{right}} - n^{\text{left}}\}, \qquad (8.15)$$

where λ is the wavelength of the radiation, and l is the path covered. Hence it is seen that, since $\lambda \ll l$, the rotation angle may be significant even in the case of a small difference between the refractive indices of the right and left circularly polarized light.

Now, let us combine the above arguments into a unique chain of reasoning. The presence of neutral weak currents (the existence of the neutral Z^0 intermediate boson) gives rise to an admixture of weak interaction to the electromagnetic interaction between particles in an atom. Owing to this admixture, atomic states exhibit violation of parity, and, consequently, the rigorous selection rules based on parity are removed in the case of electromagnetic transitions in atoms. Spontaneous electromagnetic emission between levels with violated parity turns out to be circularly polarized, since the transition probabilities between such levels do not coincide for right- and left-polarized light; the circular polarization P of the spontaneous radiation may serve as a measure of parity violation in the atom. Another aspect of the same phenomenon consists in optical activity originating in atomic medium: parity violation in atoms is displayed as rotation of the plane of polarization of radiation passing through such media (which is devoid of optical activity in the usual sense of this term).

8.2 Experimental Observation of Parity Violation in Atoms

Although the example considered in Fig. 8.1 is convenient for revealing the physical nature of the effect, it is extremely unsatisfactory from the point of view that, here, weak interaction supplements an allowed $E1$ transition with an $M1$ transition having an intensity that in atoms is always much lower than the intensity of the $E1$ transition. From (8.13) it is seen that in this case the expected circular polarization is particularly small, not only because of the mixing coefficient β being small, but also owing to the relationship between the matrix elements of the allowed and forbidden transitions being disadvantageous from the point of view of the parity nonconservation effect. For achieving a measurable P value it would be more favorable if the situation were opposite and a weak $M1$ transition were supplemented with a strong $E1$ transition.

Searches for a favorable combination of conditions for the observation of parity violation in atoms were under way in the 1970s in several laboratories of different countries. What were these conditions? First of all, it was necessary to find states exhibiting a mixing relative to parity that was as high as possible. Such states are most probable in the region of heavy atoms, since the mixing coefficient β grows rapidly with the atomic number Z (approximately like Z^3). Further, it had to be a case of strongly suppressed $M1$ or $E2$ transition, so that even a small admixture of the $E1$ transition could result in a noticeable difference between the probabilities of transitions excited by right- and left-polarized light, respectively. However, the stronger the suppression of the principal mode of electromagnetic transition between a certain pair of levels of the atom, the narrower the corresponding line is in the emission or absorption spectrum of the atom considered. This meant that the beam of light impinging upon the target had to be both extremely intense and extremely monochromatic. Such requirements could be met by a laser beam.

Parity violation in atoms was discovered in 1978 in the Novosibirsk Institute of Nuclear Physics of the Siberian Branch of the USSR Academy of Sciences. Rotation of the plane of polarization of light was observed for a laser beam passing through bismuth vapour.

Figure 8.2 shows the pattern of the lowest levels of the bismuth atom. The principal configuration of this atom is $6s^2 6p^3$. All the levels of this configuration, from the ground level $6s^2 6p^3 : {}^4S_{3/2}$ up to the level $6s^2 6p^3 : {}^2P_{3/2}$, have negative parity, and, consequently, electromagnetic transitions between

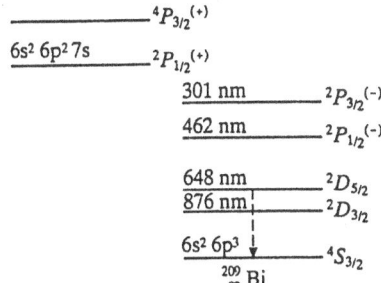

Fig. 8.2. Disposition of the lowest levels in the bismuth atom

them are either $M1$ or $E2$ transitions. Among these it is possible to single out the $^2D_{5/2} \rightarrow {}^4S_{3/2}$ transition, to which corresponds the known red line of wavelength $\lambda = 648\,\text{nm}$. According to the usual selection rules, this is a typical mixed $M1 + E2$ transition. However, in this case the $M1$ transition and the $E2$ transition are both strongly suppressed, since they both proceed only because of the admixture of other configurations (of the same parity) with the $^2D_{5/2}$ and $^4S_{3/2}$ states.

During preparation of the Novosibirsk experiment, various theoretical groups performed thorough calculations of the wave functions, of the probabilities of electromagnetic transitions, and of the circular polarization P of the emission for many transitions in a whole series of atoms that were considered to be candidates for the experiment on parity violation. The $^2D_{5/2} \rightarrow {}^4S_{3/2}$ transition in bismuth turned out to be one of the most suitable transitions:

$$P(\text{Bi})|_{\text{theor}} = 3.8 \times 10^{-7}. \tag{8.16}$$

This was one of the largest values expected for the circular polarization of emission in atoms. The carefully performed experiment yielded a value that was in excellent agreement with this theoretical prediction:

$$P(\text{Bi})|_{\text{exp}} = (4.04 \pm 0.54) \times 10^{-7}. \tag{8.17}$$

The observation of parity violation in atoms confirmed the validity of the fundamental claims of weak interaction theory concerning neutral currents and turned out to be one of the most important steps towards the discovery of intermediate bosons.

Ninth Lecture

9.1 Electromagnetic Radiation Caused by the Passage of Particles Through Matter: Direct Processes

In this lecture we shall examine the generation mechanisms and properties of electromagnetic radiation induced by particles passing through matter. This is a very broad topic; we shall consider only electromagnetic radiation associated with the passage of charged particles, heavy and light, through matter at low and high energies and under conditions in which the atoms of the medium are arranged orderly and disorderly. We shall first deal with direct photon generation processes and then with processes related either to the excitation of particles passing through matter, or to the excitation of the atoms and molecules of the medium itself, de-excitation of which gives rise to electromagnetic radiation.

We have already become familiar with a process of the first kind in Lecture 4: such was radiative stopping of electrons, a process determining their energy losses at energies exceeding the critical energy. But therein we were interested in only the motion of the radiating electron itself and set aside everything related to the radiation generated. However, both sides of the phenomenon of radiative stopping are quite clear if viewed from the same theoretical standpoint, and quantum electrodynamics, from which in Sect. 4.5 we took the expression for specific radiative losses, $(-dE/dx)_{rad}$, also provides expressions for the energy spectrum and angular distribution of bremsstrahlung photons. When $E \gg m_e c^2$, the probability of an electron of energy E emitting a photon with an energy in the $(E', E' + dE')$ interval per unit path of the electron is calculated using the formula

$$W(E, E')dE' = \frac{1}{x_0} \frac{dE'}{E'} , \tag{9.1}$$

which works quite well everywhere except at the very edges of the spectrum: $E' \to 0$ and $E' \to E'_{max} = E$. The quantity x_0 occurring in (9.1) is the radiation length (4.19). From (9.1) it is seen that the distribution of the number of photons decreases monotonically, whereas the bremsstrahlung energy distribution is uniform:

$$E'W(E, E')dE' = \frac{1}{x_0} dE' . \tag{9.2}$$

Note that all the photons of energies between zero and $\frac{1}{2}E$ generated by an electron passing through a certain layer Δx carry away exactly the same total energy as all the photons of energies from $\frac{1}{2}E$ up to the maximum value of E.

A more precise formula for the bremsstrahlung spectrum is given by the Schiff formula derived in the lowest order of perturbation theory within the framework of quantum electrodynamics and by applying the Thomas–Fermi model in accounting for screening of the atomic nucleus by the electrons. We shall not present this formula here owing to its being too cumbersome, but we shall demonstrate the shape of the Schiff spectrum in Fig. 9.1. As we see, the bremsstrahlung energy distribution slowly becomes uniform as the electron energy increases.

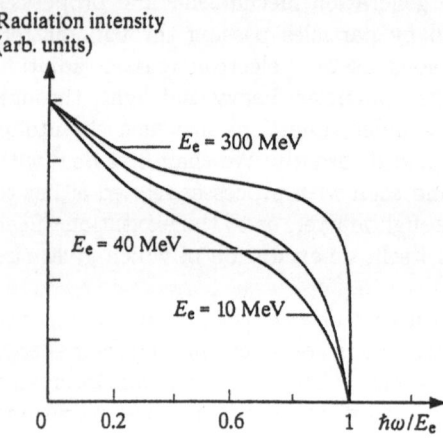

Fig. 9.1. Bremsstrahlung spectrum of electrons of various energies in platinum ($Z = 78$)

Bremsstrahlung is characterized by its angular distribution being sharply forward directed; the mean angle of the outgoing photons is estimated by the formula

$$\bar{\vartheta} \approx \frac{m_e c^2}{E}. \tag{9.3}$$

A particular case of bremsstrahlung is *synchrotron radiation* (SR), although it is not actually related to the passage of particles through matter, but arises from the gyration of particles in a uniform magnetic field. It is also termed *magnetic bremsstrahlung*. The greater the Lorentz factor $\gamma = E/mc^2$ of a particle, the more powerful the synchrotron radiation it causes. SR, like bremsstrahlung of particles in matter, is most significant in the case of electrons and positrons. In the ultrarelativistic limit of $\gamma \gg 1$ synchrotron radiation is concentrated in the plane of the particle's orbit and is sharply directed along its instantaneous velocity vector. The opening angle of the SR cone is calculated by the same formula, (9.3), as used for the opening angle

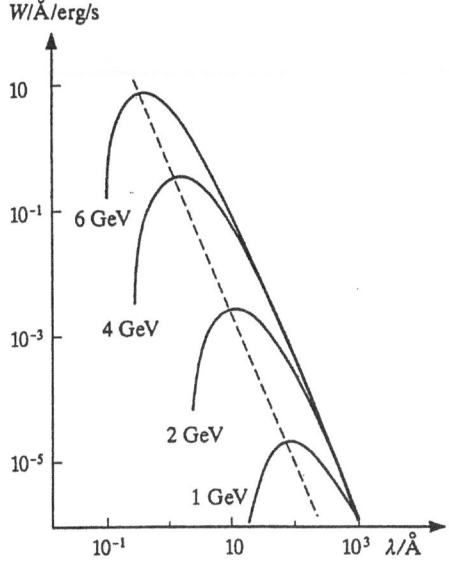

W/Å/erg/s

10

10^{-1} 6 GeV

10^{-3} 4 GeV

2 GeV

10^{-5}

1 GeV

10^{-1} 10 10^3 λ/Å

Fig. 9.2. Spectral distribution of synchrotron radiation for various electron energies

of the bremsstrahlung cone: $\overline{\vartheta} = m_e c^2 / E$. The SR spectrum is continuous (Fig. 9.2), and maximum intensity is observed at the frequency

$$\nu_{\max} = \frac{eH_\perp}{m_e c} \gamma^2, \tag{9.4}$$

where H is the strength of the magnetic field in which the orbit of the particle lies. Synchrotron radiation is polarized: when SR is observed in the plane of its orbit, the electric field strength vector is characterized by a very large component parallel to this plane and by a very small component perpendicular to it. Under such observation conditions the synchrotron radiation of ultrarelativistic electrons and positrons is practically linearly polarized.

In a very strong inhomogeneous magnetic field, an electron may move in a special manner, being apparently "tied" to the line of force of the field: its trajectory looks like a very narrow spiral stretched out along this line of force, while the particle moves along the line of force itself and follows all its twists. The magnetic bremsstrahlung of such electrons (it is called the *bending radiation*) is characterized by a series of special properties. The issue of the bending radiation of electrons and positrons in the vicinity of a pulsar occupies an important place in astrophysics.

Now, let us once again turn to the radiation of particles in matter. If the velocity v of a particle in a medium with a refractive index n exceeds the speed of propagation of electromagnetic waves in it,

$$v > \frac{c}{n}, \tag{9.5}$$

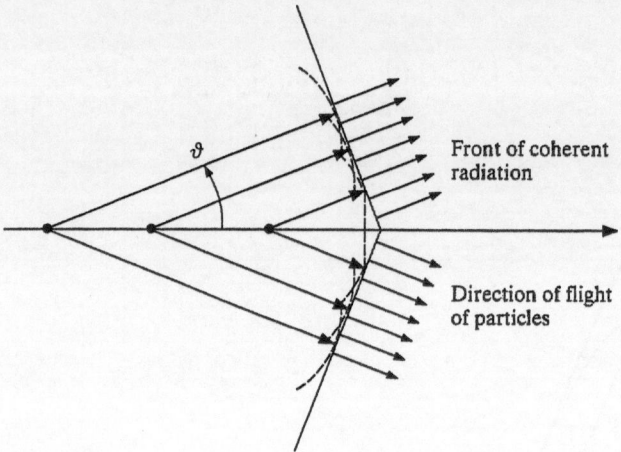

Fig. 9.3. Construction of the Cherenkov radiation front with the aid of Huygens principle

then the radiation of separate dipoles excited in the medium by the electric field of the particle adds up to form a characteristic coherent radiation propagating in the wake of the particle passing by. This is the Vavilov–Cherenkov radiation (*Cherenkov radiation*) (Fig. 9.3). The opening angle of the cone, along the generatrices of which the Cherenkov radiation propagates, is determined by the relation

$$\cos\vartheta = \frac{c/n}{v} = \frac{1}{\beta n} \leq 1 \,, \tag{9.6}$$

i. e., it is larger, the greater the velocity of the particle is. At the minimum (threshold) velocity $v_{\text{thresh}} = c/n$, in the given medium, for which Cherenkov radiation can still be generated, this cone degenerates into a narrow beam along the direction of motion of the particle. Table 9.1 presents threshold values of the Lorentz factor of a particle, $\gamma_{\text{thresh}} = E_{\text{thresh}}/mc^2 = (1-1/n^2)^{-1/2}$, which is the point from which the generation of Cherenkov radiation becomes possible.

The photon spectrum of Cherenkov radiation is calculated by the formula

$$N(h\nu)\mathrm{d}(h\nu) = (2\pi)^2 \frac{Z^2 e^2}{hc^2}\left(1 - \frac{1}{n(\nu)^2\rho^2}\right)\mathrm{d}(h\nu) \,. \tag{9.7}$$

Table 9.1. Threshold for the generation of Cherenkov radiation, γ_{thresh}, in some gaseous media

He	Ne	Air
120	85	40

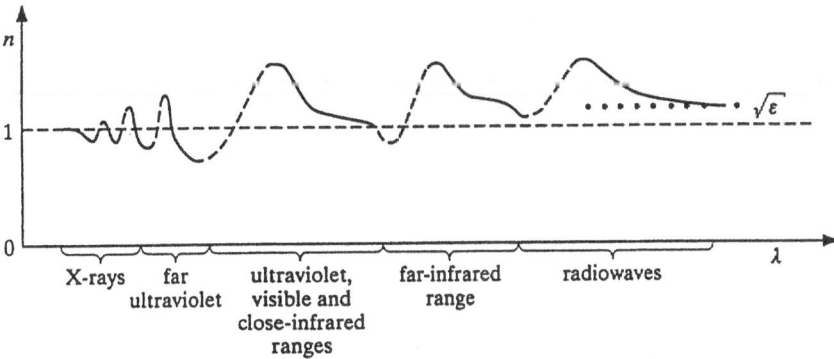

Fig. 9.4. Dependence of the index of refraction n of transparent media on the radiation wavelength

If one forgets that the index of refraction of the medium, $n(\nu)$, occurring in (9.7) depends on the frequency of the electromagnetic radiation, the radiation spectrum may seem to be uniform and the main radiation energy may seem to be concentrated in the region of the highest frequencies. Actually, no Cherenkov radiation at all exists in the X-ray region or in the far ultraviolet region, where $n < 1$ (Fig. 9.4). From the part at high frequencies, we see that the spectrum of Cherenkov radiation practically terminates in the close ultraviolet region. The uniformity of the spectrum (9.7) within the visible range results in the intensity of its short-wave (blue–violet) edge being essentially higher than the intensity of its long-wave (red) edge (hence the characteristic bluish colour of Cherenkov radiation).

Close in its physical nature to Cherenkov radiation is the transition radiation arising when a charged particle crosses the border between two media with differing refractive indices. Transition radiation propagates in the two directions from the partition boundary. The spectral region of backward-directed radiation is narrow (mainly the visible range), and that of forward-directed radiation is, on the contrary, very broad, and in this case the maximum energy of the emitted photons is, with high precision, proportional to the particle energy; at electron energies of several GeV it amounts to several keV.

9.2 Characteristic Radiation of the Atoms of a Medium Due to the Interaction of Particles with Matter

If a vacancy is produced in an inner shell of an atom of a medium when a passing particle ionizes the atom, then this atom turns out to be in a highly excited state, and occupation of the vacancy by an electron from a higher shell results in the atom releasing significant energy. This energy is

released in discrete portions, which represent the differences in energies between levels of the atomic electron, $\varepsilon_i \to \varepsilon_j$, and is carried away either by
another electron knocked out from a more weakly bound external shell (the
so-called *radiationless*, or *Auger*, *transition*), or by a photon produced directly
in the course of a vacancy being filled (fluorescence). Table 9.2 presents the
standard notation for the fluorescence lines corresponding to occupation of
vacancies in the K and L shells of atoms (also see Table C.2); for most atoms
these transitions happen to be in the X-ray region, hence the second term:
X-ray lines, or *characteristic X-ray radiation*. The notation presented is applied not only for X-ray lines, but also for lines of the Auger spectrum. For
this reason Table 9.2 includes both optically allowed and optically forbidden
transitions (K_{α_3} and others). Some of the Auger transitions, when the electron moves from one subshell to another within the same shell (for example,
$2p_{3/2} \to 2p_{1/2}$, $2s_{1/2}$; $2p_{1/2} \to 2s_{1/2}$), are sometimes classified as a special
group. Such are the so-called Koster–Cronig transitions. They differ from
other Auger transitions by having lower energies and higher rates (the latter
is related to the wave functions of atomic electrons strongly overlapping in
calculations of transition matrix elements when these electrons are in the
same shell).

Table 9.2. X-ray lines of atoms (K and L series)

Transition	Standard notation for the transition	Standard notation for the line
$2p_{3/2} \to 1s_{1/2}$	$K - L_{\mathrm{III}}$	K_{α_1}
$2p_{1/2} \to 1s_{1/2}$	$K - L_{\mathrm{II}}$	K_{α_2}
$2s_{1/2} \to 1s_{1/2}$	$K - L_{\mathrm{I}}$	K_{α_3}
$3p_{3/2} \to 1s_{1/2}$	$K - M_{\mathrm{III}}$	K_{β_1}
$3p_{1/2} \to 1s_{1/2}$	$K - M_{\mathrm{II}}$	K_{β_2}
$3d_{5/2} \to 2p_{3/2}$	$L_{\mathrm{III}} - M_{\mathrm{V}}$	L_{α_1}
$3d_{3/2} \to 2p_{3/2}$	$L_{\mathrm{III}} - M_{\mathrm{IV}}$	L_{α_2}
$3d_{3/2} \to 2p_{1/2}$	$L_{\mathrm{II}} - M_{\mathrm{IV}}$	L_{β_1}

Thus, the yield of characteristic radiation or of Auger electrons due to the
interaction of particles with matter points to the formation of vacancies in
the inner shells of atoms of the medium; a quantitative measurement of this
yield permits us to obtain the partial ionization cross sections for individual
electron shells of the atoms. Figure 9.5 presents the integral ionization cross
section of the M shell of the xenon atom, measured in this way.

Data on the dependence of the yields of individual characteristic lines on
the impact parameter of charged particles incident upon atoms of a medium
are of significant interest. If we could vary the collision parameter we could
direct the incident beam toward the locations of different shells: toward the
most strongly bound shells in the case of central collisions, and toward the

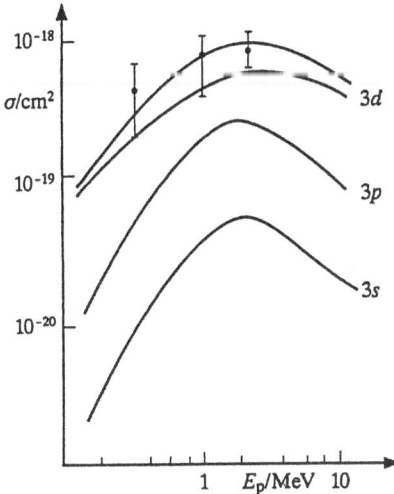

Fig. 9.5. Ionization cross section for the M shell of a xenon atom; the curves are calculated theoretically and the contributions of subshells $3s$, $3p$, and $3d$ are shown [9.1]. $E_{\hat{p}}$ is the energy of the incident protons

most weakly bound shells in the case of peripheral collisions. The correspondence between the yield of characteristic radiation and the impact parameter can be established in correlation experiments, in which the charged particle scattered by an atom of the medium is detected at a certain scattering angle in coincidence with this radiation (Fig. 9.6). Knowing the interaction potential of the incident particle with an atom of the medium, $V(r)$, it is possible, by applying the laws of classical mechanics, to relate the impact parameter b to the scattering angle of the particle, ϑ. In the center-of-mass system of the colliding particles this relationship is given by the formula $\vartheta_{\mathrm{cm}} = \pi - 2\varphi_0$, where

$$\varphi_0 = \int_{r_{\min}}^{\infty} \frac{b}{\sqrt{1 - b^2/r^2 - V(r)/E_{\mathrm{cm}}}} \frac{\mathrm{d}r}{r} \; ; \qquad (9.8)$$

here $r_{\min} = r_{\min}(E, b)$ is the minimum distance at which the particles approach each other, and E_{cm} is the energy of their relative motion at infinity.

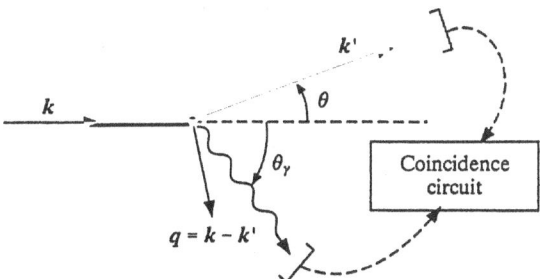

Fig. 9.6. Layout of a correlation experiment for determining the dependence of the characteristic radiation yield on the impact collision parameter

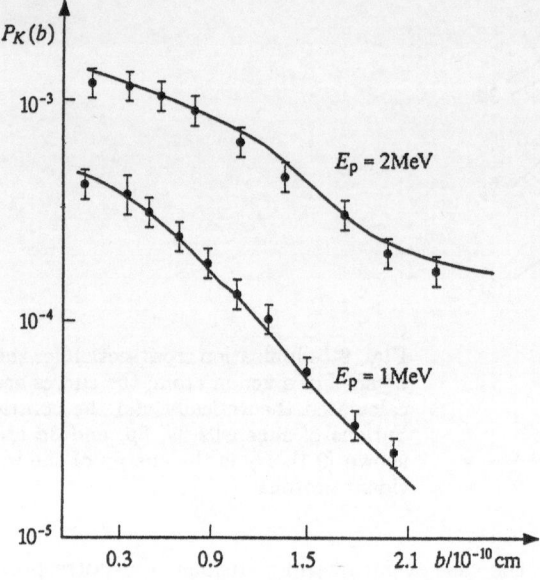

Fig. 9.7. Dependence on the impact parameter b of the ionization probability $P_K(b)$ of the K shell of copper atoms by protons [9.2]. The *solid lines* are the result of theoretical calculations

Figure 9.7 presents data on the dependence upon the impact parameter of the probability of ionization, $P_K(b)$, of the K shell of a copper atom by protons: it falls monotonically as the impact parameter increases, and within a range of several MeV it rises with the proton energy. In other cases the dependence $P(b)$ turns out to be not so simple: when nodes are present in the radial wave function of the atomic electron, oscillations may be observed in the behavior of function $P(b)$.

9.3 Angular Anisotropy of the Characteristic Radiation. The Alignment Phenomenon of Atoms in a Medium

Experiments reveal that in certain cases the angular distribution of the characteristic electromagnetic radiation of the atoms of a medium induced by a beam of particles incident upon the target turns out to be anisotropic. Under conditions in which the particles transmitted through the target are not detected, the only direction singled out in space is the direction along the incident beam. In accordance with general symmetry laws, the angular distribution of the radiation in such conditions is always symmetric with respect to rotations about this singled-out direction and, also, relative to reflection in the plane perpendicular to this direction. Consequently, it can be represented in the form of the expansion

$$W(\vartheta_\gamma) = \sum_{L=0,\,2,\,...,\,L_{\max}} a_L P_L(\cos\vartheta_\gamma), \tag{9.9}$$

where ϑ_γ is the angle between the incident beam of particles and the direction toward the detector of electromagnetic radiation, and $P_L(\cos\vartheta_\gamma)$ are Legendro polynomials of even order. In optically allowed transitions $l_{\max} \leq 2$, i.e., the shape of the angular distribution of the photons is characterized by a sole parameter:

$$W(\vartheta_\gamma) \sim 1 + \alpha_2 P_2(\cos\vartheta_\gamma). \tag{9.10}$$

When the characteristic radiation exhibits angular anisotropy, it signifies that the emitter atom somehow "remembers" the direction of the beam of particles having interacted with it. What is the mechanism allowing an individual atom of the medium to retain such information?

We shall first consider this question in a special case, taking advantage of the example of excitation of the $^1S \rightarrow {}^1P$ resonance transition in the atoms of some inert gas by a beam of fast charged particles. Let the characteristic radiation, resulting from the inverse spontaneous transition of an excited atom to the ground state, $^1P \rightarrow {}^1S$, be detected in coincidence with the scattered particle, so that each emission of a photon by an excited atom turns out to have corresponding to it a certain value of the scattered particle momentum vector. Such a principle underlies the $(e, e'\gamma)$ correlation method, one of the most informative methods of modern atomic spectroscopy (implementation of the same principle in experiments with beams of heavy charged particles involves considerable technical difficulties owing to the scattering angles of the outgoing particles lying within a very narrow angular range).

In the example considered the excited state of an atom of the medium, ^1P, is triply degenerate. Let us introduce the appropriate notation for the three sublevels: $^1P_{M=1}$, $^1P_{M=0}$, and $^1P_{M=-1}$, where M is the projection of the orbital momentum of the atom onto the quantization axis. If calculations are performed in the Born approximation, then the excitation probability of each of the sublevels ^1P_M, under the condition that the momentum of the incident particle changes from k to k', can be found from the formula [see, for example, (2.16)]

$$\left.\frac{d\sigma}{d\Omega}\right|_{{}^1S \rightarrow {}^1P_M} = \frac{k'}{k}\left(\frac{d\sigma}{d\Omega}\right)_R \left|\left\langle {}^1P_M \left| \sum_{f=1}^{Z} e^{i(k-k')\cdot r} \right| {}^1S \right\rangle\right|^2, \tag{9.11}$$

where $(d\sigma/d\Omega)_R$ is the differential cross section of Rutherford scattering from a point charge. Hence it is also possible, if necessary, to calculate the total excitation probability of the ^1P state irrespective of the orientation of the angular momentum of the atom:

$$\left.\frac{d\sigma}{d\Omega}\right|_{{}^1S \rightarrow {}^1P} = \sum_{M=0;\pm 1} \left.\frac{d\sigma}{d\Omega}\right|_{{}^1S \rightarrow {}^1P_M}. \tag{9.12}$$

We, however, are interested precisely in the orientation of this momentum. From (9.11) it is seen that, generally speaking, inelastic scattering of a charged particle results in ^1P_M states with different projections of the momentum being populated differently (nonuniformly). We shall choose the quantization axis to be directed along the momentum transfer $q = k - k'$. In this case the transition matrix elements assume the form

$$\left\langle {}^1P_M \left| \sum_{j=1}^{Z} e^{iq \cdot r_j} \right| {}^1S \right\rangle \rightarrow \left\langle {}^1P_M \left| \sum_{j=1}^{Z} e^{iqz_j} \right| {}^1S \right\rangle, \tag{9.13}$$

where, now, M is the projection of the angular momentum onto the vector q, and z_j is the projection onto this direction of the radius vector of the atomic electron. The operator $\sum_j e^{iqz_j}$ contains no azimuthal variables and, consequently, does not change the magnetic quantum numbers of the states of the atom. Therefore, of the three ${}^1P_M$ states, only the sole ${}^1P_{M=0}$ state with a zero momentum projection onto the momentum transfer $q = k - k'$ becomes populated in the considered process of inelastic scattering of a fast charged particle (in which case the Born approximation can be applied). Thus, the subsequent spontaneous electromagnetic ${}^1P \rightarrow {}^1S$ transition occurs in conditions of nonuniform population of various magnetic sublevels of the excited state, which is precisely the immediate cause of the angular anisotropy of the electromagnetic radiation in the reference system given by the vectors k and k', i.e., the cause of angular correlation between the inelastically scattered particle and the emitted photon.

The specific selection rules for the magnetic quantum number in the case of excitation of an atom by a fast particle, which were established on the basis of a quantum-mechanical analysis of this process, can also be understood with the aid of classical mechanics. Here, the analog of the Born approximation is the approximation of a single collision between a particle and an atom. In this case, the vector of the angular momentum transferred to the atom, $L = r \times p$, is strictly perpendicular to the momentum transfer vector, i.e., its projection onto the vector of the momentum transferred is zero. In essence, this is actually the selection rule $\Delta M = 0$ (when the quantization axis is directed along the vector $q = k - k'$), established above.

The special property described above of the angular momentum state of an excited atom, populated in the scattering process of fast particles, is called *alignment*. We also say, for example, that in the (e, e'γ) process the excited atom turns out to be *aligned* relative to the vector $q = k - k'$. The alignment phenomenon pertains to polarization phenomena and is peculiar to a large variety of atomic and nuclear processes. For greater clarity we point to the parallel between aligned and polarized systems. In both cases it is possible to indicate in space a certain singled-out direction, which serves as a natural quantization axis for the angular momentum of the system. The angular momentum vector $\langle L \rangle$ of a polarized system points along this singled-out direction. The mean angular momentum vector of an aligned system is zero, but (unlike in the case of a polarized system) this is achieved not by uniform population of all the magnetic sublevels, but by symmetric population of a pair of sublevels with each of the values M and $-M$ (Fig. 9.8).

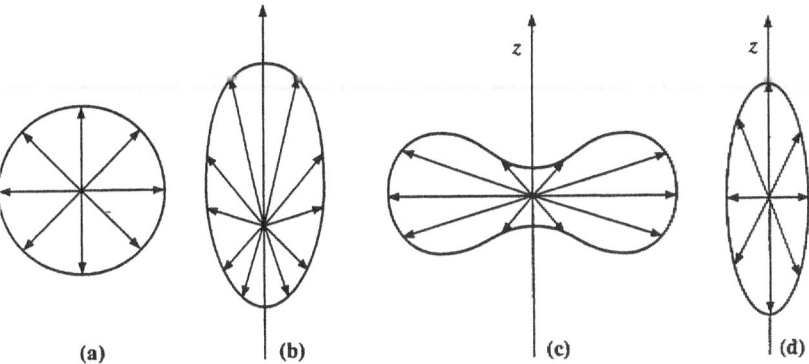

Fig. 9.8a–d. Diagram showing orientation of angular momentum of a non-polarized (**a**), a polarized (**b**), and aligned (**c**, **d**) systems

In making use of the example of correlation experiments of the $(e, e'\gamma)$ type for examining the mechanism of angular anisotropy we have digressed somewhat from our main issue of the angular anisotropy of the characteristic radiation accompanying the passage of particles through matter and detected in the simplest manner – without coincidence with the scattered particle. In these conditions the experimenter observes the integral effect corresponding to all possible directions of the scattered particle and, consequently, to the entire available range of momentum transfers. Anisotropy of the radiation in experimental conditions without coincidences indicates that such integration does not result in the alignment effect of the atomic angular momentum becoming totally blurred, vanishing, although the effect here is significantly smaller, as a rule, than that in correlation experiments. Naturally, the alignment axis now turns out to be the only remaining direction singled out in space – the direction of the incident beam. Here we note that a peculiarity of the electromagnetic radiation of an aligned system consists not only in the anisotropy of the angular distribution of the photons emitted, but also in the linear polarization of the radiation. Owing to the common origin of these two phenomena, the coefficient of angular anisotropy and the linear polarization of the radiation, P_L, are related by a rigorous algebraic relationship. For example, in the case of the $^1P \to {}^1S$ transition it is of the form $\alpha_2 = -2P_L/(3 - P_L)$.

We shall also present the example of an expression relating the linear polarization of radiation, P_L, and the relative population probabilities of the magnetic sublevels of an excited atom. Thus, if the radiation due to the $J \to J - 1$ transition is detected at an angle of 90° to the incident particle beam, then the linear polarization of the radiation is given by the expression

$$P_L = \frac{J(J+1)\sigma - 3\sum_M M^2 \sigma_M}{J(3J-1)\sigma - \sum_M M^2 \sigma_M}, \qquad (9.14)$$

where σ_M represents the integral excitation cross sections of the $|JM\rangle$ sublevels, and σ their sum: $\sigma = \sum_M \sigma_M$. In the particular case of the excited nP state we hence have for the linear polarization of the radiation in the $nP \rightarrow n'S$ transition the following: $P_L = (\sigma_0 - \sigma_1)/(\sigma_0 - \sigma_1)$. In the case of statistical population of the sublevels $(\sigma_M = \sigma/(2J+1))$ the emission of the atom is, naturally, nonpolarized and isotropic.

From the expressions presented above it is not difficult to reveal an important regularity in the variation of the character of the alignment of atoms in the medium with the energy of the incident particles. In the limit of very high energies the particles impinging upon an atom are scattered in a very narrow cone, and in each scattering act the momentum transferred to the atom (along which the angular momentum of the atom aligns) is practically perpendicular to the incident beam. In the opposite limit of the lowest energies, at the excitation threshold of the considered atomic state, the momentum transferred coincides with the momentum of the incident particle, i.e., it is directed along the incident beam. Thus, in transition from low to high energies the character of alignment alters qualitatively, and at a certain point the alignment of the atom and, consequently, the angular anisotropy and linear polarization of the radiation disappear, upon which the sign of the effect changes to its opposite (Fig. 9.9).

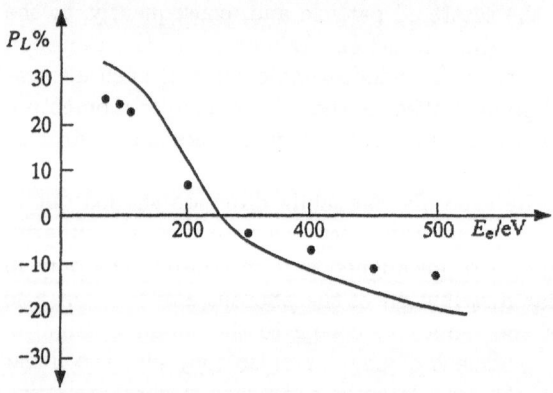

Fig. 9.9. Linear polarization P_L of the $3'D - 2'P$ line in excitation of helium atoms by electron impact [9.3]. The *solid curve* corresponds to theoretical calculation of the alignment

The mechanism of alignment of an atom under the action of a particle flux is the same when discrete (subthreshold) or autoionization (lying above the ionization threshold) states of the atom are excited. In the second case the alignment of an excited atom may be manifested, also, in the radiative and radiationless channels of its de-excitation, as an effect of angular anisotropy of the outgoing Auger electrons (Fig. 9.10).

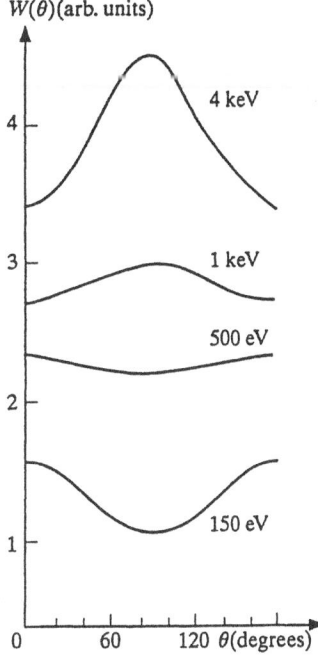

$W(\theta)$(arb. units)

Fig. 9.10. Angular distribution of Auger electrons in excitation by electron impact on the $2p^5 3s^2 : {}^2P_{3/2}$ autoionization state in sodium atoms [9.4]

9.4 Electromagnetic Radiation in Charged Particle Channelling

A spectacular manifestation of the channelling effect is presented by the orientational dependence of the yield of characteristic X-ray radiation from a medium when a monocrystalline target is bombarded with a particle beam (Fig. 9.11). The nature of the phenomenon is simple: a channelled particle interacts with the electrons in inner shells of atoms of the lattice less often than a nonchannelled particle does and, consequently, are less capable of creating vacancies in these inner shells. Another effect of the same nature is known: the output of the characteristic radiation of ions in an incident beam also exhibits a clear dip within the limits of the channelling angle.

The effect of intense radiation accompanying the channelling of light relativistic particles, predicted in the middle of the 1970s by M. A. Kumakhov, gave rise to enormous interest. The *Kumakhov radiation* has been observed and is studied in detail in many laboratories when channelling is performed of either electrons or positrons within a wide range of energies of the incident beams, from several MeV up to tens of GeV. The first experimental confirmation of the theoretical predictions was obtained in Yerevan at an electron beam energy equal to 4.7 GeV (Fig. 9.12). A characteristic feature of the Kumakhov radiation, distinguishing it qualitatively from bremsstrahlung

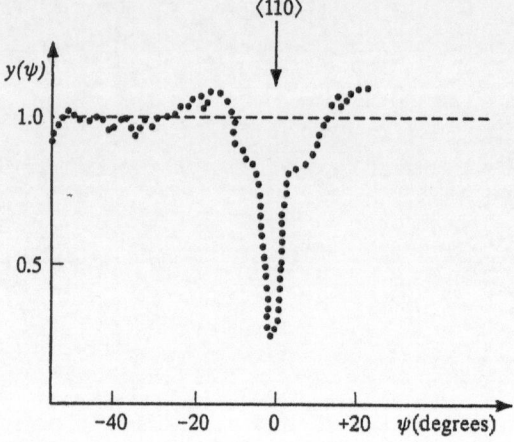

Fig. 9.11. Orientational dependence of the yield of characteristic X-ray radiation from monocrystalline copper (the L series) in the case of 200 keV incident protons; ψ is the angle between the incident beam and the $\langle 100 \rangle$ axis [9.5]

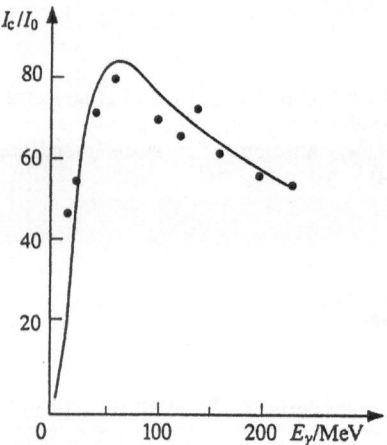

Fig. 9.12. Ratio between the yield of γ quanta, I_c, in the case of channeling along the $\langle 100 \rangle$ direction and the yield in a nonoriented target, I_0, for a 4.7 GeV electron beam inpinging upon a single diamond crystal [9.6]. The *solid curve* represents theoretical calculation in the approximation of the Lindhard continuous axial potential

and from other types of radiation, consists in the strong dependence of the intensity and shape of its spectrum on the sign of the charge of the channelled particle. This is due to the quantum character of the transverse motion of electrons and positrons in the channel: the level spectrum of a particle in the channel (transitions between these levels give rise to electromagnetic radiation) and, also, the character of how these levels are populated differ for the electrons and positrons.

Figure 9.13 shows the arrangement of the lowest levels of an electron in the axial $\langle 110 \rangle$ channel of a single silicon crystal, found theoretically by quantization of the transverse energy of the electron in the potential of a string. If this potential is approximated by the expression

$$U(\rho) = -\frac{\alpha}{\rho}, \tag{9.15}$$

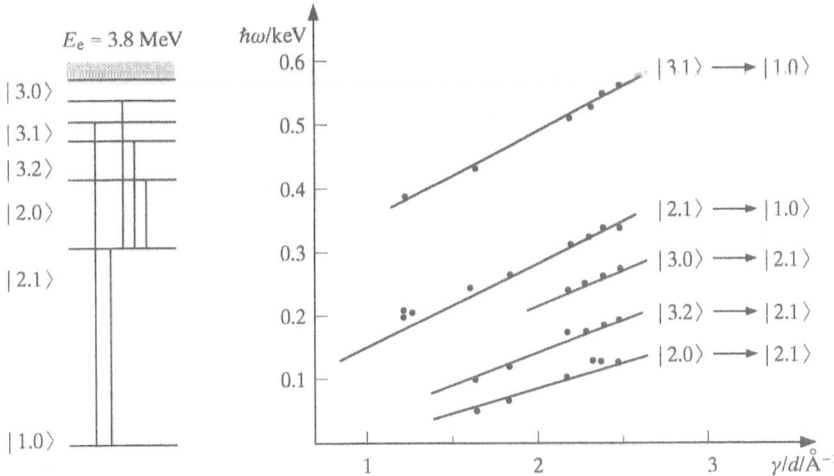

Fig. 9.13. Spectrum of levels for transverse motion of electrons in the axial $\langle 100 \rangle$ channel of monocrystalline silicon ($d = 3.84\,\text{Å}$); positions of the main peaks in the Kumakhov emission spectra versus the electron energy [9.7]

where ρ is the distance from the axis of the channel, then such a problem has an exact solution in quantum mechanics:

$$E_{\alpha,|m|} = \frac{m_e \alpha^2}{2\hbar^2 \left(n + |m| + \frac{1}{2}\right)^2} , \tag{9.16}$$

or, which is equivalent,

$$E_N = -\frac{m_e \alpha^2}{2\hbar^2 (N - 1/2)^2} . \tag{9.17}$$

Here $n = 0, 1, 2, \ldots$ is the radial quantum number, $m = 0, \pm 1, \pm 2, \ldots$ is the projection of the angular momentum of the electron onto the axis of the channel, $N = n + |m| + 1$ is the analog of the principal quantum number in the Coulomb field $V(r) = -\alpha/r$. From (9.16) it is seen that the particle levels in the potential (9.15) are characterized by random degeneracy; by taking into account deviation of the continuous potential of a string from $1/\rho$ we remove this degeneracy, which is reflected in Fig. 9.13. Among the transitions of the electron between $|N, |m|\rangle$ states, electric dipole transitions, in which the projection of the momentum, m, changes by unity, are dominant. The energy of the emitted photon in the reference frame moving at a velocity coinciding with the velocity $v_\parallel$ of forward motion of the electron along the channel axis, is simply the difference between the electron levels $E_{N,|m|}$:

$$(\hbar\omega)_{\text{mov}} = E_{N_i,|m_i|} - E_{N_f,|m_f|} . \tag{9.18}$$

In order to use the laboratory reference system we take advantage of the Lorentz transformation

$$\hbar\omega = (\hbar\omega)_{\text{mov}} \frac{1 + \frac{v_{\parallel}}{c}\cos\vartheta_{\text{mov}}}{\sqrt{1 - \left(\frac{v_{\parallel}}{c}\right)^2}}, \tag{9.19}$$

where ϑ_{mov} is the angle between the direction of the outgoing photon and the channel axis (in the moving reference system). Hence it is seen that the shape of the emitted photon spectrum repeats the shape of the electron radiation spectrum in the two-dimensional (axial-symmetric) potential $U(\rho)$, only here it is "elongated", roughly speaking, by a factor γ, where

$$\gamma = \frac{1}{\sqrt{1 - \left(\frac{v_{\parallel}}{c}\right)^2}} \tag{9.20}$$

is the Lorentz factor corresponding to forward motion of the electron. (Moreover, owing to the angular spread of the beam incident upon the crystal, each line of radiation acquires a long smoothly falling tail in the range of the lower frequencies.) The situation is as if the position of the levels corresponding to transverse motion of the electron in the channel depended, via the Lorentz factor of the electron, on its total energy E. Figure 9.13 presents the position of the main peaks in the spectra of Kumakhov radiation, measured when a 4.5 MeV electron beam bombarded a single silicon crystal. A quasiresonance structure of electron radiation spectra is also observed in the case of plane channelling (Fig. 9.14).

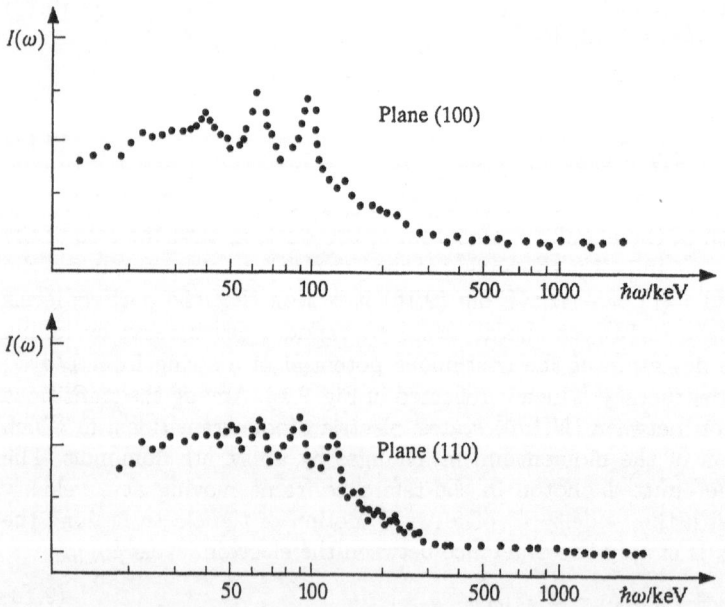

Fig. 9.14. Emission spectrum in the case of plane channeling of 56 MeV electrons in monocrystalline silicon [9.8]

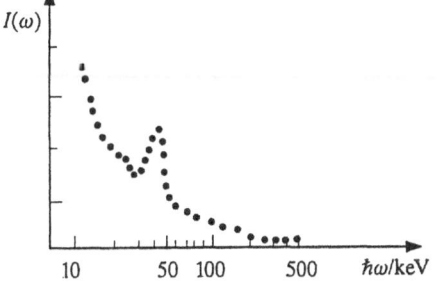

Fig. 9.15. Emission spectrum in the case of 56 MeV positrons channelled in silicon (the (110) plane channel) [9.9]

The shape of the radiation spectra when positrons are channelled is totally different (Fig. 9.15). Unlike that of electrons, the potential of transverse positron oscillations is close to being harmonic, and, consequently, the energy spectrum of transverse motion is nearly equidistant. In this case, radiation arises only in the case of transitions between neighboring levels and the spread between energies of different transitions is not large: they all blend into one, quite broad line. Typical radiation spectra in the case of the channelling of ultrarelativistic positrons are shown in Fig. 9.16.

Thus, the channelling of electrons and positrons represents a source of hard, sharply oriented electromagnetic radiation with frequencies ranging from the X-ray region up to hundreds of MeV, if modern accelerators are utilized. Further experimental and theoretical investigation of the properties and generation mechanism of this radiation (including issues such as polarization of the radiation, the influence of thermal oscillations of a crystal on its parameters, and others) promise interesting possibilities of practical applications, including those required in studies of the channelling itself as a physical phenomenon.

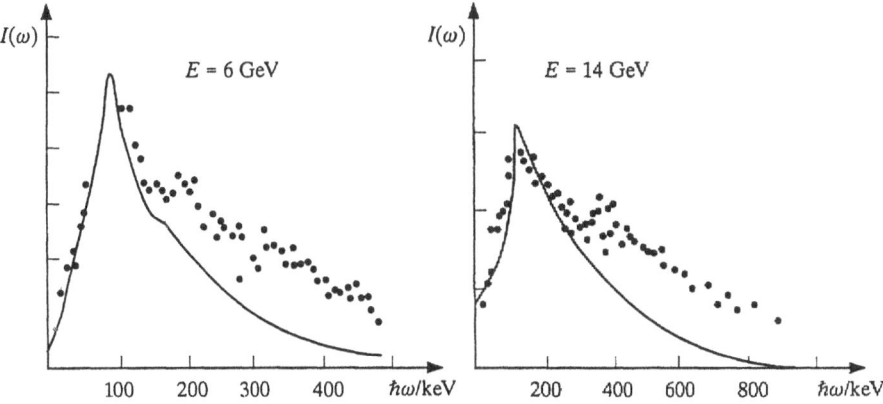

Fig. 9.16. Emission spectrum in the case positrons channeled in diamond (the (110) plane channel) [9.10]

Part III

Interaction of Neutrons with Matter

Tenth Lecture

10.1 Elementary Theory
of the Slowing Down of Neutrons

The main interaction mechanisms of neutrons with matter are elastic scattering from nuclei, inelastic scattering, the capture of neutrons by nuclei, accompanied by the emission of γ quanta (radiative capture) or fission, and other nuclear reactions [for instance, (n, p), (n, α), $(n, 2n)$]. All these processes are due to the strong interaction. Unlike most of the processes of the interaction of charged particles with the atoms of matter considered in Part I, they proceed inside a very compact central region of the atom, in which the neutron approaches the nucleus at a very close distance. The neutron has no electric charge, but in certain problems the electromagnetic interaction of the neutron with matter, due to the neutron exhibiting a magnetic moment, is of great interest.

In the elementary theory of neutron stopping we neglect all their interaction processes except nuclear elastic scattering. In each act of elastic scattering $n + A \rightarrow n + A$ the kinetic energy of the incident neutron, E_0, is shared between the energy of the scattered neutron, E, and the energy of the recoil nucleus, E_A; here E assumes the smaller part of E_0, the lighter the nucleus A and the larger the neutron scattering angle ϑ. We shall interpret the stopping process as a set of successive collisions between the neutron and the nucleus, which are independent of each other and each of which complies with the probability laws of binary collisions. Ultimately, we will be interested in the mean number of neutron collisions $\bar{n}(E_0, E)$ resulting in the energy of the neutron decreasing from the given value E_0 down to another given value E and, also, in the slow-down time and in the mean distance from the source covered by the neutron in this time.

We shall start by considering the kinematics of the elementary act of elastic scattering. It is convenient, here, to take advantage of vector diagrams, clearly showing the relations between the momenta of the colliding particles and the scattering angles both in the laboratory and in the center-of-mass reference systems (Fig. 10.1):

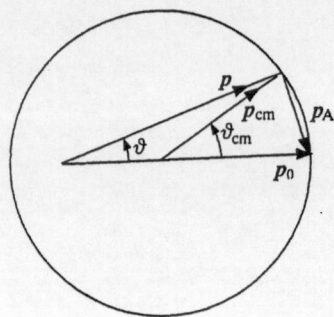

Fig. 10.1. Vector diagram for momenta in the n + A → n + A process

$$\begin{cases} p_0 = p + p_A, \\ \\ E_0 = E + E_A, \end{cases} \qquad (p_0)_{cm} = \frac{A}{A+1} p_0. \qquad (10.1)$$

Vector diagrams facilitate the finding of the energy of the scattered neutron as a function of the scattering angle:

$$E = \frac{p^2}{2m_n} = E(\vartheta). \qquad (10.2)$$

The angular distribution of neutrons scattered in elementary acts of elastic collision is given by the differential cross section $(d\sigma/d\Omega)(\vartheta)$. Knowing from kinematics the relationship between E and ϑ, it is possible to derive the energy distribution of the neutrons from their angular distribution:

$$\frac{d\sigma}{dE} = \frac{d\sigma}{d\Omega} \frac{2\pi d(\cos\vartheta)}{dE}. \qquad (10.3)$$

We stress that the first factor in this expression reflects the law of interaction between the neutron and the nucleus under consideration, while the second factor is a purely kinematical factor and is calculated with the aid of relation (10.2).

We shall further continue our computations for the particular case of neutron scattering on hydrogen nuclei, i.e., protons: n + p → n + p; all the kinematical relations are considerably simplified in this case by the masses of the colliding particles being identical (Fig. 10.2):

$$\begin{cases} E_n = E = E_0 \cos^2\vartheta, \\ \\ E_p = E_0 \sin^2\vartheta, \end{cases} \qquad \vartheta = \tfrac{1}{2}\vartheta_{cm}. \qquad (10.4)$$

Formula (10.3) contains the neutron scattering cross section in the laboratory reference system. We shall express it via the differential scattering cross section in the center-of-mass system:

$$\frac{d\sigma}{d\Omega} = \left(\frac{d\sigma}{d\Omega}\right)_{cm} \frac{d\Omega_{cm}}{d\Omega} = \left(\frac{d\sigma}{d\Omega}\right)_{cm} \frac{\sin\vartheta_{cm} d\vartheta_{cm}}{\sin\vartheta d\vartheta} = \left(\frac{d\sigma}{d\Omega}\right)_{cm} 4\cos\vartheta. \qquad (10.5)$$

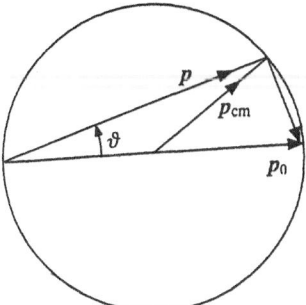

Fig. 10.2. Vector diagram for momenta in the $n + p \to n + p$ process

On the other hand, in the case of np scattering we have from (10.4)

$$\frac{\mathrm{d}E}{\mathrm{d}(\cos\vartheta)} = 2E_0 \cos\vartheta \,. \tag{10.6}$$

Substituting (10.5) and (10.6) into (10.3) we obtain

$$\frac{\mathrm{d}\sigma}{\mathrm{d}E} = \left(\frac{\mathrm{d}\sigma}{\mathrm{d}\Omega}\right)_{\mathrm{cm}} 4\pi \frac{1}{E_0} \,; \tag{10.7}$$

hence it is seen that the energy distribution of scattered neutrons (in the laboratory system) is fully determined by the angular dependence of np scattering in the center-of-mass system, i. e., by the differential cross section $(\mathrm{d}\sigma/\mathrm{d}\Omega)_{\mathrm{cm}}$. Here, it is necessary to turn to quantum theory.

According to quantum collision theory, if the de Broglie neutron wavelength exceeds the dimensions of the interaction region significantly,

$$\lambda_{\mathrm{n}} \gg R \,, \tag{10.8}$$

then the s wave ($l = 0$) is dominant in the scattering and the angular distribution of the scattered particles (in the center-of-mass system) is practically isotropic, while the differential cross section is expressed with good accuracy in terms of a single parameter – the scattering phase of the s wave:

$$\left(\frac{\mathrm{d}\sigma}{\mathrm{d}\Omega}\right)_{\mathrm{cm}} = \frac{1}{k_{\mathrm{cm}}^2} \sin^2 \delta_0(E_{\mathrm{cm}}) \,. \tag{10.9}$$

Let us check how inequality (10.8) is satisfied, if the neutron stopping process is considered, starting from a neutron energy of 1 MeV. In the center-of-mass system this energy will be 0.5 MeV. The reduced mass of the np system $\mu = 1/2m_{\mathrm{n}}$. For the de Broglie wavelength we obtain

$$\lambda = \frac{1}{k_{\mathrm{n}}} = \frac{1}{\sqrt{2E_{\mathrm{cm}}\mu}} \approx 5 \times 10^{-2} \frac{1}{\mathrm{MeV}/c} \approx 10 \,\mathrm{fm} \,. \tag{10.10}$$

This is indeed much larger, than the np interaction radius, which amounts to approximately 1 fm. We have taken $E_0 = 1\,\mathrm{MeV}$; at lower neutron energies relation (10.8) is satisfied even better. In the limit of very small energies the phase of the s wave is expressed through the only energy-independent parameter, the scattering length a, so the differential and total scattering cross sections in the center-of-mass system have quite a simple form:

$$\left(\frac{d\sigma}{d\Omega}\right)_{cm} = a^2; \qquad \sigma = 4\pi a^2. \tag{10.11}$$

Now we substitute (10.11) into (10.7):

$$\frac{d\sigma}{dE} = 4\pi a^2 \frac{1}{E_0}. \tag{10.12}$$

We see that if the angular distribution of np scattering in the center-of-mass system is isotropic, then the energy distribution of scattered neutrons in the laboratory system is uniform.

For describing the shape of the energy distribution of scattered neutrons we shall introduce the concept of a distribution function. Let $f_n(E_0, E)$ be the energy E distribution of neutrons after n elastic scattering events under the condition that the neutron had an initial energy of E_0 when undergoing the first collision. From (10.12), for function $f_1(E_0, E)$ we have

$$f_1(E_0, E) = \frac{1}{E_0}; \qquad E \le E_0. \tag{10.13}$$

In the general case $f_n(E_0, E)$ satisfies the obvious recurrent relationship

$$f_n(E_0, E) = \int_E^{E_0} f_{n-1}(E_0, E) \frac{dE'}{E'}. \tag{10.14}$$

Hence, together with (10.13), we obtain

$$f_n(E_0, E) = \frac{1}{E_0} \frac{1}{(n-1)!} \left(\ln \frac{E_0}{E}\right)^{n-1}. \tag{10.15}$$

All the $f_n(E_0, E)$ are normalized to unity:

$$\int_0^{E_0} f_n(E_0, E) dE = 1. \tag{10.16}$$

Here, the mean neutron energy after n collisions with protons is given by the simple expression

$$(\overline{E})_n = \int_0^{E_0} E f_n(E_0, E) dE = \frac{1}{2^n} E_0, \tag{10.17}$$

i.e., each collision of a neutron with a proton results in the mean energy of the neutron being halved; as we see, the stopping of neutrons in a hydrogen-containing medium proceeds very rapidly (Fig. 10.3).

Let us read dependence (10.15) in another way: what is the relative probability for a neutron to have undergone the transition $E_0 \rightarrow E$ after experiencing a certain number, n, of collisions? It is given by the expression

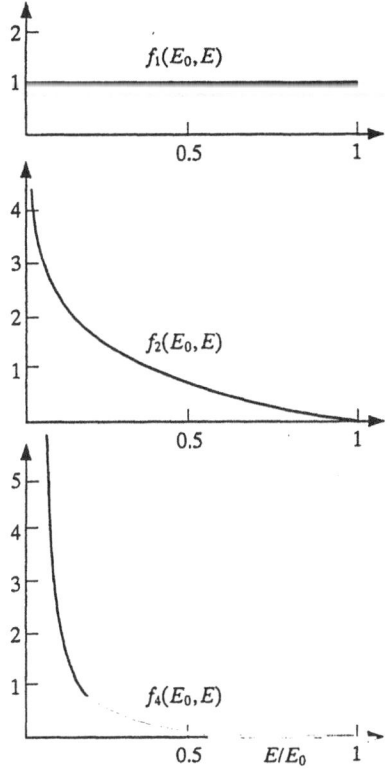

Fig. 10.3. Energy distribution $f_n(E_0, E)$ of neutrons after the first, second, and fourth collisions

$$\omega_n(E_0, E) = \frac{f_n(E_0, E)}{\sum_{n=1}^{\infty} f_n(E_0, E)} \,. \tag{10.18}$$

Hence we obtain the mean number of collisions that result in the neutron energy changing from E_0 to E:

$$\bar{n}(E_0, E) = \frac{\sum_{n=1}^{\infty} n f_n(E_0, E)}{\sum_{n=1}^{\infty} f_n(E_0, E)} = \ln \frac{E_0}{E} + 1 \,. \tag{10.19}$$

Within what intervals of E_0 and E values can this quantity be of interest to us? E_0 values of about 1 MeV are of practical interest: such, for example, is the initial energy of neutrons in a reactor (the energy of neutrons due to the fission of ^{235}U amounts to 1.46 MeV). Another, lower, boundary of the (E_0, E) interval is determined by the transition of the stopping regime in which the main part is played by successive elastic scattering of neutrons on the nuclei of the medium, to the thermalization regime when the energy exchange between the neutrons and nuclei of the medium proceeds in both directions. The thermalization process results in the mean neutron energy being of the order of the thermal energy, and the velocity distribution turns out to be of the Maxwellian type (although, strictly speaking, it differs essentially from the Maxwellian distribution, owing to capture processes). Thus, the lower boundary of the (E_0, E) interval can set to be of the order of the mean energy of the thermal motion of particles at room temperature, i.e., $\approx 0.025 \, \text{eV}$.

Quantitative characteristics of the stopping process are conventionally related to the $(1 \, \text{MeV}; 1 \, \text{eV})$ interval. At any rate, if one turns to (10.19), then it can be seen that the mean numbers of collisions within this and within the $(1 \, \text{MeV}; 0.025 \, \text{eV})$ interval do not differ significantly from each other, owing to the logarithmic energy dependence of $\bar{n}(E_0, E)$: $\ln(1 \, \text{MeV}/1 \, \text{eV}) \approx 14$; $\ln(1 \, \text{MeV}/0.025 \, \text{eV}) \approx 18$. In both cases the mean number of collisions exceeds ten, and therefore we shall omit the last term in (10.19) and apply only the simpler formula

$$\bar{n}(E_0, E) \approx \ln \frac{E_0}{E} . \tag{10.20}$$

We recall that starting with (10.4) we passed from the general case to considering the particular case of neutrons slowing down in hydrogen. Repeating all the computations performed above, it is possible to demonstrate that in the general case the mean number of collisions is given by the formula

$$\bar{n}(E_0, E) = \frac{1}{\xi} \ln \frac{E_0}{E} . \tag{10.21}$$

In the limit of $A \gg 1$ and at $A = 1$ there exist simple expressions for the kinematic factor $\xi = \xi(A)$:

$$\xi = \begin{cases} 1; & A = 1; \\ \approx \frac{2}{A}; & A \gg 1. \end{cases} \tag{10.22}$$

The slowing down of neutrons in heavy media is longer compared with how it proceeds in hydrogen, and the mean number of collisions $\bar{n}(E_0, E)$ rises proportionally to A, in accordance with (10.21).

10.2 The Mean Stopping Time of a Neutron: The Stopping Path

The elementary theory of stopping presented above permits the calculation of the mean time $\bar{t}(E_0, E)$ required for the neutron energy to drop from one fixed value down to another. It can be found as the integral

$$\bar{t}(E_0, E) = \int_0^t \mathrm{d}t' = \int_{E_0}^E \left(\frac{\mathrm{d}t'}{\mathrm{d}E'} \right) \mathrm{d}E' , \tag{10.23}$$

which is the mean rate of loss of neutron energy during the stopping process. Assume the neutron energy decreases from $E' + \Delta E'$ down to E' in a certain time interval $(t', t' + \Delta t')$ (Fig. 10.4). The mean number of collisions in this time interval equals

$$\overline{\Delta n}(E' + \Delta E', E') = \frac{1}{\xi} \ln \frac{E' + \Delta E'}{E'} \approx \frac{1}{\xi} \frac{\Delta E'}{E'} . \qquad (10.24)$$

The time between two successive collisions amounts, on average, to λ_s/v, where λ_s is the mean free path between two collisions and $v = \sqrt{2E/m_n}$ is the neutron velocity. Thus the mean time interval required for the energy of a neutron to be reduced from $E' + \Delta E'$ down to E' is

$$\Delta t' = \frac{\lambda_s}{v} \overline{\Delta n}(E' + \Delta E, E') = \frac{\lambda_s}{v} \frac{1}{\xi} \frac{\Delta E'}{E'} . \qquad (10.25)$$

We have thus found the mean neutron energy loss rate, $dE/dt = (v/\lambda_s)\xi E$, and can calculate the integral (10.23). If we consider the neutron free path length λ_s to be constant during the entire stopping interval (E_0, E), then integration gives

$$\bar{t}(E_0, E) = \frac{2\lambda_s}{\xi} \left(\frac{1}{v} - \frac{1}{v_0} \right) , \qquad (10.26)$$

where v_0 and v are the neutron velocities at the beginning and the end of the stopping interval. From the formula obtained it is seen that when $E_0 \gg E$ the mean stopping time depends very weakly on the initial neutron velocity (energy); the initial drop in energy occurs very rapidly when $E_0 \gg E$, and the time required for this drop to occur is negligible in comparison with the total stopping time (see Fig. 10.4).

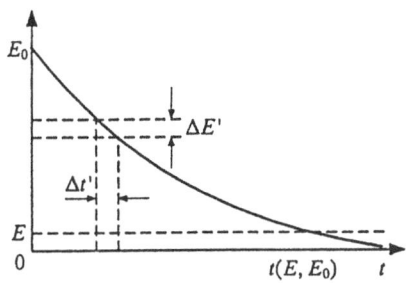

Fig. 10.4. Time diagram of neutron stopping

Now, we shall proceed to consider the spatial aspects of the stopping process. The trajectory of motion of a neutron in a medium is a tangled broken line (Fig. 10.5). Let vector r_j correspond to the segment of the trajectory between collision $(j-1)$ and collision j, so that after n collisions the neutron happens to be at the point with the coordinates

$$R_n = \sum_{j=1}^{n} r_j . \qquad (10.27)$$

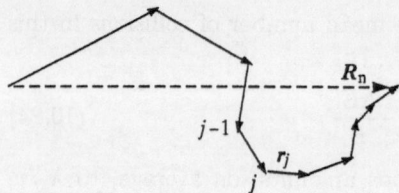

Fig. 10.5. Diagram representing the stopping of a neutron

Let us, now, calculate the mean square distance of the neutron from the starting point of its motion (i. e., from the source or the entrance point to the stopping medium):

$$\overline{R_\mathrm{n}^2} = \sum_{j=1}^{n} \overline{r_j^2} + 2 \sum_{j=1}^{n-1} \sum_{k=j+1}^{n} \overline{r_j \cdot r_k}; \qquad (10.28)$$

here, averaging is performed over a large number of independent trajectories.

The first term contains the mean square distance $\overline{r^2}$ covered by the neutron between two successive collisions. Above, we introduced the mean distance covered by a neutron between two collisions, i. e., the neutron mean free path λ_s. It would be incorrect to consider the mean square distance $\overline{r^2}$ to be simply λ_s^2. Let us find out how $\overline{r^2}$ is actually expressed through λ_s. To this end it will be convenient to consider an auxiliary problem concerning the passage of a plane-parallel neutron beam through a certain fixed layer of the material of interest. If σ_s is the scattering cross section for a neutron scattering from an atom (molecule) of the medium, and n_0 is the number of atoms (molecules) of the medium per unit volume, then the flux density of neutrons traversing a layer x without collisions is related to x by the exponential law:

$$N(x) = N_0 e^{-n_0 \sigma_\mathrm{s} x}. \qquad (10.29)$$

Knowing this law it is possible to calculate the average value of any power of the distance covered by the neutron before its first collision:

$$\overline{x^\nu} = \frac{\int s^\nu \mathrm{d}N(x)}{\int \mathrm{d}N(x)} = \frac{\nu!}{(n_0 \sigma_\mathrm{s})^\nu}. \qquad (10.30)$$

Taking into account the relationship between λ_s and σ_s,

$$\lambda_\mathrm{s} = \frac{1}{n_0 \sigma_\mathrm{s}}, \qquad (10.31)$$

we hence also obtain, together with the relation

$$\overline{x} = \lambda_\mathrm{s}, \qquad (10.32)$$

the expression sought for in our problem for the mean square distance covered before the first collision:

$$\overline{x^2} = 2\lambda_\mathrm{s}^2. \qquad (10.33)$$

Thus, the first summand in (10.28) turns out to be $2n\lambda_s^2$. Now let us calculate the second term. The scalar product $r_j \cdot r_k$ contains the cosine of the angle Θ_{jk} between the vectors r_j and r_k:

$$r_j \cdot r_k = r_j r_k \cos \Theta_{jk} . \tag{10.34}$$

To calculate the mean of such a scalar product we recall the initial physical assumption that underlies the whole consideration: the collisions of a neutron with nuclei of different atoms are independent of each other. Therefore we can write

$$\overline{r_j r_k} = \overline{r_j \, r_k} \, \overline{\cos \Theta_{jk}} = \lambda_s^2 \, \overline{\cos \Theta_{jk}} . \tag{10.35}$$

For further calculation of the mean cosine we shall take advantage of a device already applied in Lecture 1 in examining the multiple scattering of charged particles, and pass from the angle between the vectors r_j and r_k to the angle $\Theta_{j,k-1}$ between the vectors r_j and r_{k-1}:

$$\overline{\cos \Theta_{jk}} = \overline{\cos \Theta_{j,k-1}} \, \overline{\cos \vartheta_{k-1,1}} . \tag{10.36}$$

From this recurrent relation we find

$$\overline{\cos \Theta_{jk}} = (\cos \vartheta)^{k-j} , \tag{10.37}$$

where $\overline{\cos \vartheta}$ is the mean cosine of the neutron scattering angle in single collisions. Now we substitute (10.36), (10.37), and (10.33) into (10.28):

$$\overline{R_n^2} = 2n\lambda_s^2 + 2 \sum_{j=1}^{n-1} \lambda_s^2 \left| \overline{\cos \vartheta} + (\overline{\cos \vartheta})^2 + \ldots + (\overline{\cos \vartheta})^{n-j} \right| . \tag{10.38}$$

Adding up the geometric progression and further assuming $n \gg 1$ results in

$$\overline{R_n^2} \approx \frac{2n\lambda_s^2}{1 - \overline{\cos \vartheta}} . \tag{10.39}$$

Substituting here (10.21) we obtain the formula for the mean square distance from the source covered by a neutron having slowed down from the energy E_0 down to the energy E:

$$R^2(E_0, E) = 2\lambda_s \lambda_t \frac{1}{\xi} \ln \frac{E_0}{E} . \tag{10.40}$$

We have introduced a separate notation for the following combination of the mean free path and the mean cosine of the scattering angle:

$$\lambda_t \equiv \frac{\lambda_s}{1 - \overline{\cos \vartheta}} , \tag{10.41}$$

which is called the *transport length* and is widely applied not only in neutron physics, but also in many other problems related to the interaction of particles with matter. We shall examine the dependence of λ_t upon properties of the medium and on characteristics of the elementary act of elastic neutron scattering on a nucleus.

The free path length λ_s is inversely proportional to the total elastic scattering cross section,

$$\frac{1}{\lambda_s} = n_0 \sigma_s = n_0 \int \frac{d\sigma}{d\Omega} d\Omega, \tag{10.42}$$

and the transport length is related in a similar manner to another integral characteristic of the elementary act, the *transport cross section* σ_t:

$$\sigma_t = \int (1 - \cos\vartheta) \frac{d\sigma}{d\Omega} d\Omega; \tag{10.43}$$

$$\frac{1}{\lambda_t} = n_0 \sigma_s (1 - \overline{\cos\vartheta}) = n_0 \sigma_t \tag{10.44}$$

(here, $d\sigma/d\Omega$ is the differential scattering cross section of a neutron scattering from a nucleus in the laboratory system). When the neutron scattering is close to isotropic (low energies and large A), the transport cross section σ_t differs little from the total cross section σ_s, and at the same time λ_s and λ_t are close to each other. On the other hand, in those cases, when the diagram of the neutron angular distribution $d\sigma/d\Omega$ is extended in the forward direction, these relations change drastically:

$$\sigma_t \ll \sigma_s; \qquad \lambda_t \gg \lambda_s. \tag{10.45}$$

In the particular case of neutron scattering from protons we have, in accordance with (10.11) and (10.5), $\overline{\cos\vartheta} = 2/3$ and, consequently,

$$\lambda_t = 3\lambda_s. \tag{10.46}$$

A generally adopted characteristic of neutron moderators is the *stopping length* L_s:

$$L_s \equiv \sqrt{\frac{1}{6}\overline{R^2}(E_0, E)} = \sqrt{\frac{1}{3}\lambda_s \lambda_t \frac{1}{\xi} \ln \frac{E_0}{E}}, \tag{10.47}$$

where the interval (1 MeV; 0.025 eV), or sometimes (1 MeV; 1 eV), is chosen as the (E_0, E) interval. Let us estimate the stopping length of neutrons in water. Since the main contribution to stopping in H_2O is due to hydrogen, we shall neglect the contribution of oxygen, and in (10.47) we shall set $\xi = 1$ and $\lambda_t = 3\lambda_s$. We shall calculate the neutron free path via the density of water, the cross section of np scattering, the Avogadro number, and the molecular weight of water:

$$\lambda_s(H_2O) = \left[\frac{N_A \rho(H_2O)}{18} 2\sigma_s(H) \right]^{-1}. \tag{10.48}$$

The total scattering cross section $\sigma_s(H)$ for slow neutrons amounts to approximately 20 barn $= 2 \times 10^{-23}$ cm^2. We substitute this value into (10.48) and further substitute $\lambda_s(H_2O)$ into (10.47) to obtain

$$\lambda_s(H_2O) \approx 0.75\,\text{cm}\,,$$
$$L_s(H_2O) \approx 3.2\,\text{cm}\,.$$
$$(10.49)$$

What we have obtained here for L_s with the aid of simple estimations is in quite good agreement with the measured value $L_s(H_2O) = 5.4\,\text{cm}$. The reason for why our value of L_s is lower than the experimental value is that at the very beginning of the stopping process the total cross section of np scattering is noticeably lower than 20 barn, which was adopted for the estimation in the entire interval from 1 MeV down to 0.025 eV (Fig. 10.6). At the very earliest stopping stages the neutron covers larger paths between collisions and, consequently, travels a greater distance from the source than that obtained in our calculations.

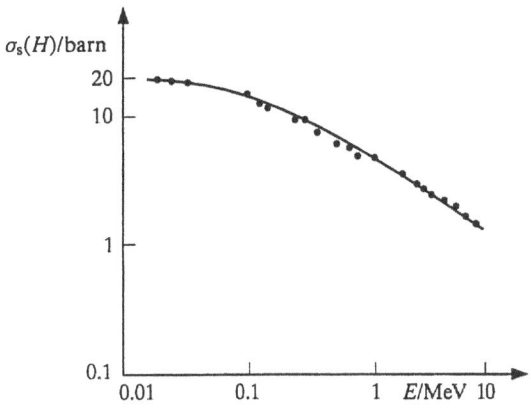

Fig. 10.6. Total elastic scattering cross section for neutrons scattering from protons

In Table 10.1 neutron stopping parameters are presented for water, heavy water, and graphite. From the point of view of all these parameters, ordinary water, H_2O, could be the most appropriate medium for utilization as a moderator in power plants involving neutrons. However, the powerful radiative capture of neutrons by protons, $n + p \rightarrow d + \gamma$, the onset of which occurs at the end of the stopping stage, considerably reduces the advantages of ordinary water.

Table 10.1. Parameters for the slowing down of neutrons in water, heavy water, and graphite

Substance	ξ	Mean free path λ_s, cm	Slowing-down time t_s, s	Slowing-down path L_s, cm
H_2O	1	1.1	1×10^{-6}	5.4
D_2O	0.73	2.6	4.6×10^{-5}	11
Graphite	0.16	2.6	1.5×10^{-4}	17.7

Eleventh Lecture

11.1 The Motion of Neutrons Upon Thermalization: The Diffusion Coefficient

Let us keep track of neutrons in a medium after their stopping and thermalization are complete. At this stage all inhomogeneity in their spatial distribution is levelled out by diffusion. In this lecture we will approach the equation of neutron diffusion within the framework of the consistent description of the motion of particle fluxes in matter, provided by physical kinetics. The main concept of physical kinetics is the distribution function. We shall reveal the properties of the neutron distribution function $f(v, r, t)$, where v is the neutron velocity in the vicinity of point r at time t, through its integral characteristics:

(a) the neutron density in space –

$$n(r, t) = \int f(v, r, t) \, \mathrm{d}^3 v \, ; \tag{11.1}$$

(b) the neutron current density –

$$j(r, t) = \int v f(v, r, t) \, \mathrm{d}^3 v \, . \tag{11.2}$$

The distribution function satisfies the Boltzmann equation, the main equation of the kinetic theory of gases and of other similar systems:

$$\frac{\partial f(v, r, t)}{\partial t} = -v \frac{\partial f}{\partial r} - \frac{F}{m} \frac{\partial f}{\partial v} + I \, . \tag{11.3}$$

The first term in the right-hand part corresponds to variation in the number of neutrons in the vicinity of the point (v, r) in phase space due to the inhomogeneity of the distribution function in space. The vector F occurring in the next summand represents the force field acting on the particles of the "gas" being considered. The last term is called the collision integral. It actually represents the difference of two integrals, which we shall write down in a conventional, symbolic form:

$$I = \int_{v' \to v} (\text{inflow}) - \int_{v \to v'} (\text{loss}) \, . \tag{11.4}$$

The first of the integrals takes into account all the collisions of a neutron with nuclei of the medium resulting in its velocity being v; the second represents an integral over all possible values of the final velocity v' given the initial neutron velocity v. Our task consists in expressing the neutron diffusion coefficient in terms of parameters of the single neutron collision with nuclei of the medium.

To this end we shall adopt the following simplifications:

(a) we shall consider the neutron distribution to be stationary,

$$\frac{\partial f}{\partial t} = 0 ; \tag{11.5}$$

(b) we exclude all external forces;
(c) we shall assume that the collision of a neutron with a nucleus does not result in a change in its kinetic energy, i.e., only the direction of its motion changes, not the absolute value of the neutron velocity v;
(d) we shall consider diffusion in a sole direction (let it be along the x axis).

Under these simplifications the Boltzmann equation assumes the form

$$v_x \frac{\partial f}{\partial x} = I , \qquad f = f(v, \alpha, x) , \tag{11.6}$$

where, now, the distribution function depends only on the neutron kinetic energy $E = mv^2/2$, the angle α between the velocity vector v and the x axis (the problem is axial symmetric), and the coordinate x.

Now we shall calculate the first and second terms in the collision integral (11.4). Under the simplifications made, integration over v' merely reduces to integration over the direction of vector v' with respect to vector v. Let ϑ be the angle between these two directions, i.e., the neutron scattering angle in an elementary act of a collision with a nucleus, and $d\sigma(v, \vartheta)/d\Omega$ the differential scattering cross section. Then, the inflow of neutrons to the vicinity of the point (v, r) in phase space and their loss are calculated by the formulae

$$\int_{v' \to v} (\text{inflow}) = vN \int \frac{d\sigma(v, \vartheta)}{d\Omega} f(v, \alpha', x) d\Omega , \tag{11.7}$$

$$\int_{v \to v'} (\text{loss}) = vN \int \frac{d\sigma(v, \vartheta)}{d\Omega} f(v, \alpha, x) d\Omega , \tag{11.8}$$

where N is the density of atoms in the medium. The resulting Boltzmann equation assumes the form

$$v_x \frac{\partial f(v, \alpha, x)}{\partial x} = vN \int \frac{d\sigma(v, \vartheta)}{d\Omega} \Big[f(v, \alpha', x) - f(v, \alpha, x) \Big] d\Omega . \tag{11.9}$$

We shall solve this integro-differential equation by the Laplace method, i.e., in the approximation in which $f(v, \alpha, x)$ is a weakly anisotropic function.

This means that, in expanding the distribution function, which depends on the angle α, as the complete set of Legendre polynomials,

$$f(v, \alpha, x) = \sum_{n=0}^{\infty} a_n P_n(\cos \alpha), \qquad (11.10)$$

it is possible to retain only the first and second terms:

$$\begin{aligned} f(v, \alpha, x) &\approx f_0(v, x) + v \cos \alpha \, f_1(v, x) \\ &= f_0(v, x) + v_x f_1(v, x). \end{aligned} \qquad (11.11)$$

In the zero approximation, in which $f = f_0(v, x)$, (11.9) assumes an extremely simple form,

$$\frac{\partial f_0}{\partial x} = 0, \qquad (11.12)$$

and there is no diffusion, so the neutron distribution is homogeneous in space:

$$n(r) = \text{const}; \quad f_0 = f_0(v). \qquad (11.13)$$

A particular case of such a distribution is the Maxwellian velocity distribution of neutrons in a uniformly heated spatially homogeneous medium:

$$f_0(v) = n_0 \left(\frac{m}{2\pi k T} \right)^{3/2} \exp\left(-\frac{mv_x^2}{2kT} \right). \qquad (11.14)$$

We shall now return to the Boltzmann equation (11.9). The physical meaning of the approximation (11.11) underlying the solution of this equation by the Laplace method is that the ordered forward motion (along the x axis) of a neutron flux described by the term $v_x f_1(v, x)$ is superimposed, like a weak background, on the random thermal motion of the neutrons. In other words, the mean velocity of the whole flux (along the x axis) is considered to be much smaller than the average speed of thermal motion of individual neutrons:

$$\left| \bar{v}_x \right| \ll v_{\text{therm}}. \qquad (11.15)$$

Therefore, upon substituting the approximate expression $f = f_0 + v_x f_1$ into (11.9),

$$v_x \frac{\partial}{\partial x} \left[f_0 + v_x f_1(v, x) \right] = vN \int \frac{d\sigma(v, \vartheta)}{d\Omega} \left[v_x' - v_x \right] f_1(v, x) d\Omega, \qquad (11.16)$$

we drop the term quadratic in v_x. Ultimately, we obtain

$$\cos \alpha \frac{\partial f_0(v, x)}{\partial x} = vN f_1(v, x) \int \frac{d\sigma(v, \vartheta)}{d\Omega} (\cos \alpha' - \cos \alpha) d\Omega. \qquad (11.17)$$

Now, we express the angle α' via the angle α, the polar scattering angle ϑ, and the azimuthal angle φ:

$$\cos \alpha' = \cos \alpha \cos \vartheta + \sin \alpha \sin \vartheta \cos \varphi. \qquad (11.18)$$

Substituting this expression into (11.17) and integrating over the azimuthal angle φ we obtain

$$\frac{\partial f_0(v,x)}{\partial x} = -vNf_1(v,x)\int(1-\cos\vartheta)\frac{d\sigma(v,\vartheta)}{d\Omega}d\Omega. \tag{11.19}$$

The right-hand part contains the neutron transport scattering cross section (10.43):

$$\sigma_t(v) = \int(1-\cos\vartheta)\frac{d\sigma(v,\vartheta)}{d\Omega}d\Omega. \tag{11.20}$$

From (11.19) it can be seen that the angular anisotropy of the neutron velocity distribution is directly related to how nonuniform in space (along the x direction) the main component $f_0(v,x)$ of the distribution function is:

$$f_1(v,x) = -\frac{1}{\sigma_t(v)Nv}\frac{\partial f_0(v,x)}{\partial x}. \tag{11.21}$$

We shall now calculate the mean neutron current density $j(x)$, in accordance with the procedure (11.2); in our case it is, naturally, directed along the x axis:

$$\begin{aligned}
j_x(x) &= \int v_x\Big[f_0(v,x)+v_xf_1(v,x)\Big]d^3v \\
&= \int v_x^2 f_1(v,x)\,d^3v = \frac{1}{3}\int v^2 f_1(v,x)\,d^3v \\
&= -\frac{1}{3}\int\frac{1}{\sigma_t(v)N}v\frac{\partial f_0(v,x)}{\partial x}d^3v.
\end{aligned} \tag{11.22}$$

Instead of the transport cross section, one can insert the transport length $\lambda_t = 1/\sigma_t N$:

$$j_x(x) = -\frac{1}{3}\int\lambda_t v\frac{\partial f_0(v,x)}{\partial x}d^3v. \tag{11.23}$$

Finally, if one assumes function $f_0(v,x)$ to factorize (i.e., that it can be represented as the product of two independent multipliers, one of which is the mean neutron density, while the other is the neutron velocity distribution):

$$f_0(v,x) = n(x)\omega(v), \tag{11.24}$$

then the mean neutron flux is given by the expression

$$j_x(x) = -\frac{1}{3}\overline{v\lambda_t}\frac{\partial n(x)}{\partial x}. \tag{11.25}$$

We have derived formula (11.25) for the particular case of neutron diffusion proceeding only in one direction. Its extension to the general case is obvious:

$$j(r) = -\frac{1}{3}\overline{v\lambda_t}\,\mathrm{grad}\,n(r). \tag{11.26}$$

In the last two expressions $\overline{v\lambda_t}$ is averaged over the neutron velocity distribution:

$$\overline{v\lambda_t} = \frac{\int v\lambda_t(v)\omega(v)\,\mathrm{d}^3v}{\int \omega(v)\,\mathrm{d}^3v}\,. \tag{11.27}$$

Formula (11.26) is usually written in the form

$$j(r) = -D\operatorname{grad} n(r)\,, \tag{11.28}$$

where the coefficient D is called the *diffusion coefficient*. Thus, we have obtained the following expression for it:

$$D = \frac{1}{3}\overline{v\lambda_t}\,. \tag{11.29}$$

11.2 The Equation of Diffusion: The Mean Lifetime of the Neutron

Let $n(r,t)$ be the neutron density in space. Its variation in time and from point to point is regulated by the balance equation

$$\frac{\partial n(r,t)}{\partial t} = -\operatorname{div} j(r,t) + q(r,t) - N\sigma_{\mathrm{abs}}vn\,, \tag{11.30}$$

where $j(r,t)$ is the neutron current density. The two last terms describe, respectively, the inflow of neutrons due to distributed sources and the loss of neutrons due to nuclear absorption reactions. Accordingly, $q(r,t)$ is the source power, while σ_{abs} is the total cross section for neutron absorption by nuclei of the medium. Substituting here $j(r,t)$ from (11.28), we obtain an equation in partial derivatives for the space density $n(r,t)$:

$$\frac{\partial n}{\partial t} = D\nabla^2 n + q - N\sigma_{\mathrm{abs}}vn\,. \tag{11.31}$$

The combination of parameters, $N\sigma_{\mathrm{abs}}v$, occurring here has the dimension of inverse time. Let us introduce a new parameter:

$$T = \frac{1}{N\sigma_{\mathrm{abs}}v}\,. \tag{11.32}$$

The combination $1/N\sigma_{\mathrm{abs}}$ is the neutron mean free path length with respect to absorption (capture) by a nucleus: thus, T is the mean lifetime of a neutron relative to the absorption process. The physical meaning of parameter T becomes even more clear if the balance equation (11.31) is considered under the special condition of $D \to 0$ (extremely slow diffusion), and the source of thermal neutrons is "switched on" for only a very short time. Then, after this time the balance equation acquires the form

$$\frac{\partial n(r,t)}{\partial t} = -\frac{1}{T}n(r,t)\,, \tag{11.33}$$

and its solution is

$$n(r,t) = n(r,0)e^{-t/T}.\tag{11.34}$$

Hence, it is clearly seen that T is the mean lifetime of a neutron with respect to absorption by nuclei of the medium. We shall term T the *diffusion time*.

In the new notation the balance equation assumes its final form:

$$\frac{\partial n(r,t)}{\partial t} = D\nabla^2 n(r,t) + q(r,t) - \frac{1}{T}n(r,t).\tag{11.35}$$

We shall call it the *equation of diffusion*.

11.3 Typical Problems
of Thermal Neutron Diffusion Theory

11.3.1 A Stationary Point Source in an Infinite Medium

Consider a source of thermal neutrons of constant intensity in the vicinity of the point $r = 0$ and with dimensions that can be considered negligible. Then, everywhere except at $r = 0$ we have

$$D\Delta n - \frac{n}{T} = 0.\tag{11.36}$$

The solution of this equation in partial derivatives is well known:

$$n(r) = \text{const}\,\frac{e^{-r/L}}{r}.\tag{11.37}$$

Here, we have introduced a new notation,

$$L = \sqrt{DT},\tag{11.38}$$

which is the *diffusion length*. The parameter L shows how far from the source the neutrons are on average, as a result of diffusion. Let us calculate, making use of solution (11.37), the mean square distance of a neutron from the source:

$$\langle r^2 \rangle = \frac{\int r^2 n(r)\, d^3 r}{\int n(r)\, d^3 r} = 6L^2.\tag{11.39}$$

Thus, the root-mean-square value of the distance covered by a neutron from the source is

$$r_{\text{rms}} = \sqrt{6}\,L.\tag{11.40}$$

Now it is possible to correct the formulation of the problem discussed: the source can be considered point-like if its dimensions are much smaller than the diffusion length L.

Table 11.1 presents parameters of thermal neutron diffusion for a series of media.

Table 11.1. Diffusion parameters of thermal neutrons in water, heavy water, and graphite

Substance	Diffusion time t_s/ s	Diffusion length L_s/cm
H_2O	2×10^{-4}	2.7
D_2O	0.15	160
Graphite	1.2×10^{-2}	54

11.3.2 Boundary Conditions for Neutrons Passing from a Medium to Vacuum

This is a very important problem in the theory of neutron interaction with matter, and here we shall give the result of its approximate solution, without presenting the actual calculation.

Consider a flux of thermal neutrons leaving a region occupied by matter and entering a vacuum (Fig. 11.1). The boundary conditions at the interface surface Σ should reflect the fact that to the right of the surface the velocity vectors v of all the neutrons are directed away from it, while to the left, owing to collisions with nuclei of the medium in the boundary layer, some of the neutrons may also move in the opposite direction into the region, where diffusion takes place. The thickness of the layer where such a special diffusion mode occurs is of the order of the transport length λ_t. For many practical purposes the following device turns out to be quite suitable in solving problems of neutron diffusion within a restricted volume: the neutron density is set to zero at an imaginary surface Σ', specially introduced and shifted away from the interface Σ out into the external region at a certain distance Δ, termed the *extrapolation length:*

$$n\Big|_{\Sigma'} = 0 . \tag{11.41}$$

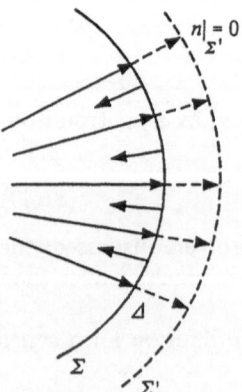

Fig. 11.1. Boundary conditions for neutron at the exit from the medium into vacuum

This distance is of the order of the transport length λ_t. Numerical calculations give good results if Δ is chosen according to the formula

$$\Delta = 0.7\lambda_t \,. \tag{11.42}$$

11.3.3 Diffusion of Thermal Neutrons from a Point-like Pulsed Source

Let the source be switched on for a very short time interval $\Delta t \ll T$; we want to find the neutron density distribution within a volume of given configuration at $t > \Delta t$. This is a typical boundary problem in mathematical physics. We shall consider its solution in the special case of a point-like pulsed source at the center of a homogeneous sphere, and of neutrons from the source distributed isotropically. When so formulated, our problem is strictly spherically symmetric, and the neutron density $n(r, t)$ at each moment in time depends only on the distance from the source: $n(r, t) \to n(r, t)$. The problem formulated reduces to solving the equation in partial derivatives for a function of two variables with the appropriate boundary condition

$$\begin{cases} \dfrac{\partial n(r,t)}{\partial t} = D\Delta n - \dfrac{n}{T}\,, \\ n(r = R, t) = 0\,; \end{cases} \tag{11.43}$$

here R is the radius of the sphere with the extrapolation length Δ taken into account.

Let us take advantage of the method of the separation of variables and look for a solution to the equation in the form

$$n(r, t) = S(r)e^{-t/T}g(t)\,, \tag{11.44}$$

and single out here the time-dependent exponential factor $e^{-t/T}$, corresponding to the mode of extremely slow diffusion (11.34). Now (11.43) will become

$$S(r)\frac{dg(t)}{dt} = D\Delta S(r)g(t)\,, \tag{11.45}$$

which we solve by introducing a parameter for the separation of variables:

$$\frac{1}{Dg(t)}\frac{dg}{dt} = \frac{\Delta S}{S} = -\Omega = \text{const}\,. \tag{11.46}$$

The radial function $S(r)$ is found from the problem of eigenvalues:

$$\begin{cases} \Delta S + \Omega S = 0\,, \\ S(r = R) = 0\,, \end{cases} \tag{11.47}$$

namely,

$$S_\nu(r) = A_\nu \frac{\sin \sqrt{\Omega_\nu}\, r}{r}\,, \tag{11.48}$$

where

$$\sqrt{\Omega_\nu} = \nu \frac{\pi}{R}, \qquad \nu = 1, 2, \ldots. \tag{11.49}$$

Here

$$g_\nu(t) = e^{-\Omega_\nu D t}. \tag{11.50}$$

Thus, the neutron density $n(r,t)$ represents a superposition of space-time spherically symmetric harmonics:

$$
\begin{aligned}
n(r,t) &= \sum_{\nu=1}^{\infty} A_\nu \frac{\sin \frac{\pi \nu}{R} r}{r} \exp\left[-\left(\nu^2 \frac{\pi^2 D}{R^2} + \frac{1}{T}\right)t\right] \\
&= \sum_{\nu=1}^{\infty} A_\nu \frac{\sin \frac{\pi \nu}{R} r}{r} \exp\left\{-\left[\nu^2 \pi^2 \left(\frac{L}{R}\right)^2 + 1\right]\frac{t}{T}\right\};
\end{aligned} \tag{11.51}
$$

the constants $A_1, A_2, \ldots$ are determined by the shape and intensity of the initial neutron pulse. Higher harmonics decay more rapidly than the main harmonic, namely in a time of the order of $T(R/L)^2 = R^2/D$. When this happens, the time dependence of the diffusion becomes exponential:

$$
n(r,t)\Big|_{t \gg R^2/D} \simeq \frac{\sin \frac{\pi r}{R}}{r} \exp\left\{-\left[\pi^2 \left(\frac{L}{R}\right)^2 + 1\right]\frac{t}{T}\right\}. \tag{11.52}
$$

Twelfth Lecture

12.1 Neutron Diffraction in a Crystal

At low neutron energies, when the neutron de Broglie wavelength is comparable with the distance between the atoms of the medium, scattered waves caused by the interaction of neutrons with the nuclei of differing atoms interfere with each other – the phenomenon of diffraction arises. We shall start to examine the nature of neutron diffraction and its main features by making use of the example of a crystal with a perfect lattice, assuming it to be motionless and all the nuclei of the atoms to be identical.

Let $v(r)$ be the potential energy of the interaction of a neutron with the nucleus of an atom. In considering a set of N such nuclei fixed at points $b_1, b_2, \ldots, b_N$ we write the potential energy of the interaction of a neutron with the whole set in the form of a sum of operators of pair interaction:

$$\hat{V}_N = \sum_{j=1}^{N} v(r - b_j), \tag{12.1}$$

where r represents the coordinates of the neutron in the reference system in which the vectors b_j are given. Our immediate task is to calculate the differential cross section of elastic scattering of a neutron from the set of nuclei at rest. We shall start with perturbation theory: the Born approximation. In this approximation the elastic scattering amplitude $F_N(k \rightarrow k')$ is a matrix element of the first order with respect to $\hat{V}_N$, where the wave functions of the incident and scattered neutron are approximated by plane waves:

$$F_N(k \rightarrow k') = -\frac{m_n}{2\pi\hbar^2} \int e^{-ik'\cdot r} \hat{V}_N(b_j, \ldots, b_N; r) e^{-ik\cdot r} \, d^3r$$

$$= -\frac{m_n}{2\pi\hbar^2} \int \sum_{j=1}^{N} v(r - b_j) e^{-iq\cdot r} \, d^3r$$

$$= \sum_{j=1}^{N} \left[-\frac{m_n}{2\pi\hbar^2} \int v(r - b_j) e^{-iq\cdot(r-b_j)} \, d^3r \right] e^{iq\cdot b_j}. \tag{12.2}$$

The expression in square brackets is nothing but the Born scattering amplitude of a neutron scattered by individual nuclei:

$$f(k \to k') = -\frac{m_n}{2\pi\hbar^2} \int v(r)e^{iq \cdot r} \, d^3r \,, \tag{12.3}$$

where the direction and absolute value of the momentum transfer vector

$$q = k - k' \tag{12.4}$$

is determined by the momenta k and k' of the incident and scattered neutron, respectively, or, which is the same, by its kinetic energy E and scattering angle ϑ.

Now we substitute (12.3) into (12.2). Then, because all the N elementary amplitudes are identical, the total scattering amplitude $F_N(k \to k')$ and the differential cross section for the whole set of N nuclei arise before us in a factorized form:

$$F_N(k \to k') = f(k \to k') \sum_{j=1}^{N} e^{iq \cdot b_j} \,; \tag{12.5}$$

$$\left(\frac{d\sigma}{d\Omega}\right)_N = |f(k \to k')|^2 \left|\sum_{j=1}^{N} e^{iq \cdot b_j}\right|^2 = \left(\frac{d\sigma}{d\Omega}\right)_0 \left|\sum_{j=1}^{N} e^{iq \cdot b_j}\right|^2, \tag{12.6}$$

where the first factor carries information on the neutron interaction with an individual nucleus and has nothing to do with the medium to which this nucleus pertains, and the second factor depends only on the order in which these N nuclei are arranged. In nuclear and atomic physics the probability of a scattering process is conventionally characterized by the effective cross section reduced to a single nucleus, a single atom, or a single molecule. In our case, we shall find the differential cross section for neutron scattering from the whole given set of nuclei reduced to a single nucleus:

$$\frac{d\sigma}{d\Omega} = \frac{1}{N}\left(\frac{d\sigma}{d\Omega}\right)_N = \left(\frac{d\sigma}{d\Omega}\right)_0 \frac{\left|\sum_{j=1}^{N} e^{iq \cdot b_j}\right|^2}{N}. \tag{12.7}$$

Here

$$\left(\frac{d\sigma}{d\Omega}\right)_0 = |f(k \to k')|^2 \tag{12.8}$$

is the differential cross section for scattering of a neutron by a sole isolated nucleus. We shall call the second factor in (12.7), carrying information on the arrangement of the N nuclei in the target, the *structure factor* (the word "structure" indicates here that it is related to the structure of the target).

Before proceeding to analyze the structure factor for the set of nuclei of interest, located at the nodes of an ideal crystalline lattice, it is necessary to say a few words concerning the Born approximation underlying the derivation of the formulae obtained. The interaction of a neutron with a nucleus is a strong interaction, and at low neutron energies the Born approximation is absolutely unsuitable for describing the scattering of neutrons from nuclei. Nevertheless, the final expression (12.7), where the cross section for scattering from a set of nuclei is expressed through the cross section for scattering from a separate nucleus, may be applied in our problem. The point is that expression (12.5), from which everything else follows, is valid, with some minor correction, also when the Born approximation does not work. This is the so-called *impulse approximation*:

$$F_N(k \to k') = \sum_{j=1}^{N} f_j(k \to k') e^{iq \cdot b_j} , \qquad (12.9)$$

which unlike (12.5), instead of Born amplitudes, contains the exact scattering amplitudes of the incident particle on the constituent elements of the target. The physical meaning of the impulse approximation is that the amplitude of scattering from a set of centers is the sum of all the elementary scattering amplitudes for each of the centers, corrected by the phase factors $e^{iq \cdot b_j}$, which depend on the mutual disposition of these centers.

We shall analyze the properties of the structure factor for the case of the simplest, cubic, lattice. Let our N nuclei form a rectangular block, where the number of rows in the x, y, and z directions are, respectively, N_x, N_y, and N_z: $N = N_x N_y N_z$. The coordinates of each of the nuclei are given by a set of integer numbers j_x, j_y, and j_z:

$$b_{jx} = b j_x; \quad j_x = 0, 1, \ldots, N_x - 1 \text{ and so on}, \qquad (12.10)$$

where b is the lattice constant. The sum of exponentials in (12.7) factorizes into the product of three sums:

$$\sum_{j=1}^{N} e^{iq \cdot b_j} = \sum_{j_x=0}^{N_x-1} e^{iq_x b j_x} \sum_{j_y=0}^{N_y-1} e^{iq_y b j_y} \sum_{j_z=0}^{N_z-1} e^{iq_z b j_z} . \qquad (12.11)$$

Let us calculate the first one:

$$\sum_{j_x=0}^{N_x-1} e^{iq_x b j_x} = \frac{1 - e^{N_x i q_x b}}{1 - e^{iq_x b}} = \pm \frac{e^{\frac{1}{2} N_x i q_x b}}{e^{\frac{1}{2} i q_x b_x}} \frac{\sin \frac{1}{2} N_x (q_x b - 2\pi \tau_x)}{\sin \frac{1}{2} (q_x b - 2\pi \tau_x)} . \qquad (12.12)$$

This relation is an identity for any integer number τ_x:

$$\tau_x = 0; \pm 1; \pm 2; \ldots . \qquad (12.13)$$

Like in formula (12.11), the structure factor in (12.7) also factorizes into three factors. We shall calculate the first one for a macroscopically large crystalline block ($N_x \to \infty$):

$$\lim_{N_x \to \infty} \frac{1}{N_x} \left| \sum_{j_x} e^{iq_x b j_x} \right|^2$$

$$= \lim_{N_x \to \infty} \frac{1}{N_x} \left(\frac{\sin \frac{1}{2} N_x b (q_x - \frac{2\pi}{b} \tau_x)}{\sin \frac{1}{2} b (q_x - \frac{2\pi}{b} \tau_x)} \right)^2 = \frac{2\pi}{b} \delta \left(q_x - \frac{2\pi}{b} \tau_x \right) . \qquad (12.14)$$

Hence we obtain for the entire structure factor

$$\lim_{N_x,N_y,N_z\to\infty} \frac{\left|\sum_{j=1}^{N} e^{iq\cdot b_j}\right|^2}{N} = \left(\frac{2\pi}{b}\right)^3 \delta(q - K).$$ (12.15)

The infinite sequence of vectors K occurring here, which are *inverse lattice vectors*, is constructed according to the rule

$$K = \left\{\frac{2\pi}{b}\tau_x, \quad \frac{2\pi}{b}\tau_y, \quad \frac{2\pi}{b}\tau_z\right\},$$ (12.16)

where τ_x, τ_y, τ_z are integer numbers.

Thus, the differential cross section for scattering of neutrons from an ideal rigid crystalline lattice reduced to a single nucleus of the target substance is given by the expression

$$\frac{d\sigma}{d\Omega} = \left(\frac{d\sigma}{d\Omega}\right)_0 \left(\frac{2\pi}{b}\right)^3 \delta(k - k' - K).$$ (12.17)

Hence it is seen that neutron scattering from an oriented single-crystal sample occurs only along certain directions determined by the condition

$$k - k' = K,$$ (12.18)

where K is one of the inverse lattice vectors of the crystal. In all other directions the mutual interference of waves due to neutron scattering from different nuclei suppresses it completely. Everything is precisely like in optics or, better to say, in the physics of X-rays, and condition (12.18) is essentially nothing but the *Wolf–Bragg reflection law* known in X-ray optics.

Consider the case of a symmetric neutron reflection from the (100) facet of a crystal with a simple cubic lattice (Fig. 12.1). The grazing angle equals the reflection angle ψ, and the scattering angle ϑ is the grazing angle doubled:

$$\vartheta = 2\psi.$$ (12.19)

In the case of such a geometry the length of the momentum transfer vector $q = k - k'$ is

$$|k - k'| = 2k\sin\frac{\vartheta}{2} = 2k\sin\psi,$$ (12.20)

and the vector itself is perpendicular to the crystalline surface. Thus, neutron scattering results in a "selection" of inverse lattice vectors K oriented along the z axis. Expressing the absolute value of the neutron momentum k (wave vector) via its de Broglie wavelength,

$$k = \frac{2\pi}{\lambda},$$ (12.21)

and taking into account that only integer numbers τ_z of the same sign correspond to reflection, we obtain from (12.18) the traditional formulation for the Wolf–Bragg law:

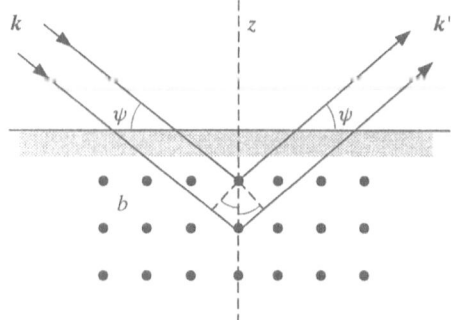

Fig. 12.1. Diagram of symmetric reflection of neutrons from a monocrystalline surface

$$n\lambda = 2b\sin\psi\,; \quad n = 1, 2, \ldots . \tag{12.22}$$

From Fig. 12.1 it is clearly seen that the right-hand part of (12.22) contains nothing but the path difference for two adjacent "rays" of the neutron wave. Interference of these rays turns out to be constructive (the scattering is coherent), when an integer number of neutron wavelengths are contained in this path difference. Precisely this fact is expressed by condition (12.22).

12.2 Coherent and Incoherent Neutron Scattering

In the preceding section we assumed that all the nuclei of a crystalline lattice are completely identical and, consequently, that the interactions of a neutron with these nuclei are also completely identical. Actually, such a rigorous homogeneity of a crystal is violated by a whole series of factors, and this introduces an element of incoherence into the scattering of neutrons.

We shall first consider this issue from a formal standpoint, without going into a discussion of the physical nature of the inhomogeneities of a crystal. Assume the scattering amplitudes of neutrons scattered from different target nuclei differ somehow from each other, for some reason. Let us write this down in terms of the initial formula (12.9):

$$f_j(k \to k') \neq f_{j'}(k \to k')\,, \qquad j \neq j'\,. \tag{12.23}$$

It is very important to assume that throughout the whole crystal there exists no correlation between the deviations of any two amplitudes from the mean value. We shall define this mean value in the usual way and consider the deviation of each amplitude from the mean:

$$\overline{f}(k \to k') = \frac{1}{N}\sum_{j=1}^{N} f_j(k \to k')\,, \tag{12.24}$$

$$\Delta f_j(k \to k') = f_j(k \to k') - \overline{f}(k \to k')\,.$$

Clearly, in this case

$$\sum_j \Delta f_j = 0. \tag{12.25}$$

Now, as we did for (12.7), we shall calculate the differential scattering cross section reduced to a single nucleus, i.e., averaged over the whole sample:

$$\frac{d\bar{\sigma}}{d\Omega} = \frac{1}{N}\left|\sum_j f_j e^{i\mathbf{q}\cdot\mathbf{b}_j}\right|^2 = |\bar{f}(\mathbf{k}\to\mathbf{k}')|^2 \frac{1}{N}\left|\sum_j e^{i\mathbf{q}\cdot\mathbf{b}_j}\right|^2$$

$$+\frac{1}{N}\sum_{j,j'=1}^{N}\Delta f_j \Delta f_{j'} e^{i\mathbf{q}\cdot(\mathbf{b}_j-\mathbf{b}_{j'})}. \tag{12.26}$$

The first factor is seen to be the structure factor of the coherent scattering cross section, which is already familiar to us. In the second summand we shall retain only the diagonal terms of the sum with $j = j'$, taking into account the chaotic character of the deviations Δf_j from zero. Thus, if uniformly distributed inhomogeneities in the crystal exist, the coherent scattering, while retaining the selectivity inherent in diffraction phenomena, relative to the direction of the outgoing neutron and to the target orientation, weakens in comparison with the case of an ideal uniform crystal and is now present together with a background of incoherent scattering, smoothly varying with the angle:

$$\frac{d\sigma}{d\Omega} = \left(\frac{d\sigma}{d\Omega}\right)_{\text{coh}} + \left(\frac{d\sigma}{d\Omega}\right)_{\text{inc}}, \tag{12.27}$$

$$\left(\frac{d\sigma}{d\Omega}\right)_{\text{coh}} = |\bar{f}(\mathbf{k}\to\mathbf{k}')|^2 \left(\frac{2\pi}{b}\right)^3 \delta(\mathbf{k}-\mathbf{k}'-\mathbf{K}), \tag{12.28}$$

$$\left(\frac{d\sigma}{d\Omega}\right)_{\text{inc}} = \frac{1}{N}\sum_{j=1}^{N}|\Delta f(\mathbf{k}\to\mathbf{k}')|^2. \tag{12.29}$$

What gives rise to incoherence in neutron diffraction in a single crystal? First of all, admixtures and mechanical defects in the lattice do. However, even in the crystal of a definite chemical element incoherence may arise owing to its isotopic inhomogeneity, since the neutron cross sections for different isotopes of the same element generally have nothing in common. Further, if the target nuclei have spin, then the interaction of neutrons with them exhibits a dependence on the mutual orientation of the spins of the neutron and nucleus. Finally, incoherence is created by chaotic thermal oscillations of the nuclei in the lattice.

12.3 The Influence of Thermal Oscillations of the Lattice: The Phenomenon of Inelastic Diffraction

It is convenient here to repeat the arguments presented in discussing the Mössbauer effect (see Sect. 7.2). The interaction of a lattice nucleus with an external perturbation (there, absorption or emission of a γ quantum; here, neutron scattering) results in momentum being transferred to the crystal. In some fraction of events it is taken up by the entire lattice. Only then do conditions arise for the resonance absorption of γ quanta (the Mössbauer effect). On the other hand, only then is elastic scattering of neutrons by the crystal possible without energy being transferred to it. In Sect. 7.2 the fraction of events involving recoilless emission or absorption of a γ quantum was termed the attenuation factor. In the Debye model this is the Debye–Waller factor (7.39). A precisely similar factor determines the fraction of neutrons scattered without recoil in their elastic (from the point of view of the elementary act $n + A \rightarrow n + A$) scattering on a crystal:

$$f_{\text{D-W}} = \exp\left\{ -\frac{3}{2} \frac{R}{\hbar\omega_{\text{D}}} \left[1 + 4\left(\frac{T}{T_{\text{D}}}\right)^2 \right] \int\limits_0^{T_{\text{D}}/T} \frac{u\,du}{e^u - 1} \right\}. \qquad (12.30)$$

The difference from formula (7.39) consists only in that here R is the recoil energy that a free nucleus would have received in the case of neutron scattering at a given angle:

$$R = \frac{\hbar^2 |\boldsymbol{k} - \boldsymbol{k}'|^2}{2M}. \qquad (12.31)$$

The Debye–Waller factor also determines the fraction of other events in which, from the point of view of the elementary act $n + A \rightarrow n + A$, the neutron is scattered elastically (i. e., without excitation or fragmentation of the nucleus), but part of its energy is spent on the excitation of nuclear oscillations in the lattice; the total fraction of all such events is $1 - f_{\text{D-W}}$. According to the Debye model, the spectrum of nuclear oscillations in a solid exhibits the universal shape (7.18), and one solid differs from another in this aspect only by its Debye frequency ω_{D}, or, which is the same, by its Debye temperature T_{D}. Actually, the spectra of lattice oscillations in different substances are very individual. Neutron diffraction involving the excitation of oscillation quanta characteristic for each particular lattice, i. e., of phonons, serves precisely as a powerful method for the investigation of solids. Such processes of inelastic diffraction are conveniently represented graphically:

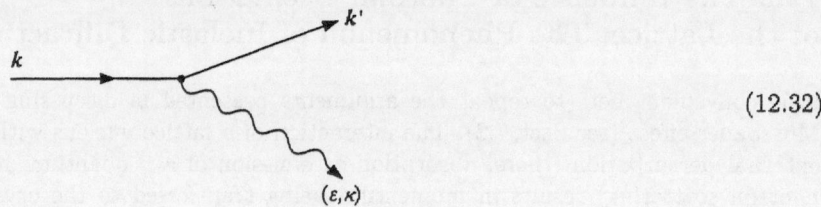

(12.32)

Here, the wavy line corresponds to a phonon carrying energy ε and momentum κ. The inverse process is also possible when neutron scattering results not in the production of a phonon, but in its absorption, i. e. energy is transferred from the crystal to the neutron incident upon it:

(12.33)

Inelastic diffraction of this type is especially important and interesting for applications of cold and ultracold neutrons for spectroscopic studies of solids.

12.4 Neutron Scattering by Polycrystals

A polycrystalline target composed of a chemically pure substance can be imagined as a very large number of closely packed small single crystals oriented chaotically in space (Fig. 12.2). The scattering of an incident neutron flux on each such small crystal proceeds in accordance with the picture shown in Sect. 12.1. Such scattering is possible only if the momentum exchanged between the neutron and the small crystal coincides with one of the inverse lattice vectors of this small crystal. Since each small crystal has corresponding to it an infinite number of such vectors, and in the sample there are very many such small crystals situated randomly with respect to the neutron beam, the Bragg peaks corresponding to neutron diffraction in separate small crystals blend into a continuous picture without any sharp bumps representing the angular distribution of scattered neutrons.

An exception occurs only for neutrons with wavelengths exceeding the doubled lattice constant:

$$\lambda > 2b . \tag{12.34}$$

In this case the momentum transferred to the lattice,

$$|k - k'| < \frac{2\pi}{b} , \tag{12.35}$$

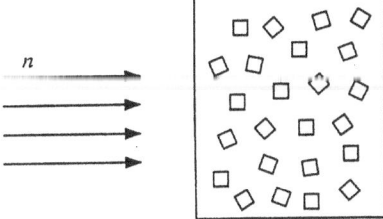

Fig. 12.2. Neutron scattering in a poly-crystalline target

is restricted to such an extent that the condition for diffraction (12.18) is not satisfied for any of the inverse lattice vectors. The diffraction of such neutrons is not possible for any orientation of the small monocrystals inside the sample, so neither is it possible for a whole polycrystalline target.

Thus, a polycrystalline sample may serve as a filter that does not transmit the entire short-wave part of the incident neutron spectrum with $\lambda < 2b$. Only neutrons with energies

$$E < E_{\text{thresh}} = \frac{\pi^2 \hbar^2}{2m_n b^2} \,. \tag{12.36}$$

pass through such a filter without being scattered. Substituting into (12.36) typical values, $b = (2\text{–}3) \times 10^{-8}$ cm, we obtain $E_{\text{thresh}} \leq 10^{-2}$ eV, which corresponds to the region of thermal neutrons.

Mesoatomic and Mesomolecular Processes

Thirteenth Lecture

13.1 Mesoatoms and Their Properties

Herein we shall return to the processes of the interaction of charged particles with matter, and we shall supplement Lectures 1–6 by considering a special class of such processes related to the formation of mesoatoms and mesomolecules and their interaction with matter. The very terms *mesoatom* (mesonic atom) and *mesomolecule* (mesonic molecule) are conventional and at present are not just used for denoting, in the strict sense of the terms, atoms and molecules in which a negatively charged meson plays the part of the orbital particle. The terms are extended to any "exotic" atomic systems that may contain a π^- or a K^- meson, a muon μ^-, an antiproton $\bar{p}$ or a Σ^- hyperon (data on the parameters of these elementary particles are presented in Tables A.2 and A.3).

The simplest of such systems is the hydrogen muonic atom H_μ. It is actually an exact copy of the hydrogen atom. The muon, not being stable, has practically no influence on the properties of the H_μ atom, since the mean lifetime of the muon exceeds by several orders of magnitude the times characteristic of the main processes proceeding in a muonic atom – the revolution period of the muon on the atomic orbit or the lifetimes of excited states of the H_μ atom relative to spontaneous radiative transitions. Like other mesoatoms, formation of the muonic atom H_μ results from atomic capture of a μ^- muon onto a high atomic orbit, from which its cascade of electromagnetic transitions to the ground state of the muonic atom, i.e., to the K-orbit, proceeds. Here it lives a long time (on time scales characteristic of mesoatomic processes), and ultimately either decays ($\mu^- \rightarrow e^- + \nu_\mu + \bar{\nu}_e$) or is absorbed by the nucleus ($p + \mu \rightarrow \nu + \nu_\mu$). This is the so-called μ capture. The last two processes compete with each other and are both due to the weak interaction.

The muon is 207 times heavier than the electron, and therefore [see (C.6)] the orbits of the H_μ atom are the same number of times smaller than the electron orbits in the hydrogen atom. On the contrary, the whole scheme of levels of the H_μ atom is 207 times more "stretched out" in comparison with the same scheme in the hydrogen atom: the ionization potential of the H_μ muonic atom in the ground state is 2.7 keV instead of 13.5 eV; the K_α line in the spectrum of its electromagnetic radiation is about 2 keV instead of 10 eV, i.e., it happens to be in the X-ray region; and so on. Investigation of the spectra of the characteristic X-ray radiation is the main method applied to keep track of the cascade of electromagnetic transitions in a mesoatom and, upon determination of the position of its energy levels, to establish the shift of these energy levels due to distortion of the purely Coulomb interaction of the muon with the nucleus at small distances. In intermediate and heavy atoms the main contribution to this effect is due to the atomic nucleus not being point-like. Let us, for example, estimate the dimensions of the K orbit of the muonic atom of uranium $(\mu^-\,_{92}\mathrm{U})$: $\langle r_\mu \rangle = a_0/Z\,(m/m_e)^{-1} \approx 2.5 \times 10^{-13}$ cm. This is less than the dimensions of the uranium nucleus itself: $R = r_0 A^{1/3}$ cm. Thus, the muon spends a significant part of its time inside the nucleus, where its electrostatic interaction with the charge of the nucleus is weakened in comparison with what it would be for a point-like charge (see Fig. C.3), and so the binding energy of the lowest levels (especially of the ns levels) of the muonic atom is lower than given by Bohr's simple law:

$$E_\mathrm{n} = -\left(\frac{m}{m_\mathrm{e}}\right)\frac{Z^2\epsilon_0}{2n^2}.\tag{13.1}$$

Thus, in the $(\mu^-\,_{92}\mathrm{U})$ muonic atom the $2p \rightarrow 1s$ transition energy, which should be over 18 MeV according to this formula, is actually 6.3 MeV (Fig. 13.1).

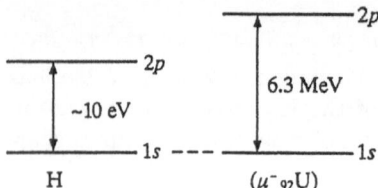

Fig. **13.1.** Comparison of the positions of the $2p$ level for an electron in a hydrogen atom and for a muon in the $(\mu^-\,_{92}\mathrm{U})$ mesonic atom

The nucleus not being point-like is not the only effect to be taken into account in precise calculations of the energy spectra of muonic atoms. Relativistic effects are strong in heavy mesoatoms; this means that the Dirac equation must be applied in calculations, instead of the Schrödinger equation. The radiative corrections are strongly enhanced as compared to those for ordinary atoms. The muon being close to the nucleus in heavy atoms creates favorable conditions for the electromagnetic field of the muon to polarize the nucleus itself; the reciprocal influence of this polarization effect on the wave functions and positions of the energy levels of the muon in the mesoatom is very significant. In nuclear physics much attention is paid to various effects due to the interaction between the intrinsic and muon degrees of freedom of the mesoatom. One such effect, the *radiationless transition* of the muon from orbit to orbit, arises when the distance between the levels of the mesoatom is close to the excitation energy of some collective level of the nucleus; in the case of such a "resonance" the cascade of radiation transitions of the muon may be terminated, and the remaining energy of the orbital motion of the muon is transferred to the nucleus.

Even more complex processes associated with the nucleus take place in hadron atoms, i.e., in pionic, kaonic, antiproton and Σ^- atoms. In going through a cascade of electromagnetic transitions in the mesoatom, a negatively charged hadron initially moves far away from the nucleus and at this stage essentially behaves like a muon. Upon reaching the orbit, where the overlapping of its wave function and dense nuclear matter is essential, the hadron "burns up", being engaged in strong interaction with the nucleons of the nucleus. In the case of the pion, this mainly results in two-nucleon absorption, $(\pi^-, 2n)$ or (π^-, np), for the kaon this is the reaction $K^-N \rightarrow \pi\Lambda$. The antiproton undergoes pair annihilation on a nucleon of the nucleus, and the energy thus released ($\sim 2\,\text{GeV}$) is carried away by the pions produced or by other, heavier, mesons. The so-called process of $\Sigma \rightarrow \Lambda$ conversion is characteristic of the Σ hyperon: $\Sigma^-N \rightarrow \Lambda N$. In hadron atoms of sufficiently high Z, the cascade of transitions is terminated at excited levels and does not reach the K orbit. This is clearly seen when X-ray spectra are deciphered. Thus, in the spectra of pionic atoms with $Z > 11$ no K_α line is present, with $Z > 30$ no L_α line is present, etc.

The main purpose of experiments with hadron atoms in nuclear physics is the investigation of hadron–nuclear interactions at low energies. Strong interaction in hadron atoms not only is manifest in the termination of the cascades of electromagnetic transitions, but also leads to a shift and broadening of the levels of these atoms. The set of data on all these effects permits the introduction of essential corrections to the optical potentials of hadron–nuclear interactions. In turn, these potentials themselves are involved in the calculations of transition cascades in hadron atoms. From such computations it is known that the competition between the absorption of a hadron by the nucleus directly from a high-lying atomic orbit and the electromagnetic transition from such an orbit to a lower orbit play an important part in the formation of the cascade at the stage preceding the arrival of the hadron at the "last orbit".

From which orbit does the cascade of mesoatomic transitions start? In other words, how are the initial populations of mesoatomic levels distributed in the process of muon or hadron capture by an atom? At present no final answer to this question exists. It is very difficult to provide a rigorous theoretical examination and quantitative description of the stage at which the mesoatom is formed, i.e., when the muon or hadron undergoes transition from the continuous spectrum to a bound atomic orbit. But the main reason is that at present this stage of mesoatomic processes still remains completely unavailable for direct experimental control.

The main role in the atomic capture of a negatively charged muon or hadron is played by the mechanism for substitution of a new particle for an atomic electron,

$$A + \mu^-(\pi^-, K^-, \ldots) \rightarrow A_{\mu^-(\pi^-, K^-, \ldots)} + e, \qquad (13.2)$$

due to the direct interaction between them in the presence of a third body – the atomic residue. From the very first publications in mesoatomic physics, the new particle was considered, in the formation of a mesoatom, to occupy an orbit at the same mean distance from the nucleus as the atomic electron taking part in the substitution process. Hence it is possible to estimate the number of the initial orbit. If one makes use of (C.7), which is valid for hydrogen-like systems, and takes into account the dependence of the atomic scale of distances on the mass of the orbital particle,

$$a_x = a_0 \left(\frac{m_x}{m_e} \right), \qquad (13.3)$$

then one finds that if the new particle substitutes an atomic K electron, then it occupies an $|i\rangle$ orbit with a principal quantum number estimated by the formula

$$n_i \approx \sqrt{m_x/m_e}, \qquad (13.4)$$

where m_x is the mass (to be more precise, the reduced mass) of the muon or hadron. In the case of a muon $n_i = 14$, for a pion $n_i = 17$, for a kaon $n_i = 31$, etc.

For a long time, these estimates were used as a direct indication of where a cascade originates. No other data existed: before meson factories were put into operation, the accuracy with which X-ray spectra of mesoatoms were measured did not permit resolution of lines corresponding to transitions from very high levels (in the case of muonic atoms, for example, from levels with

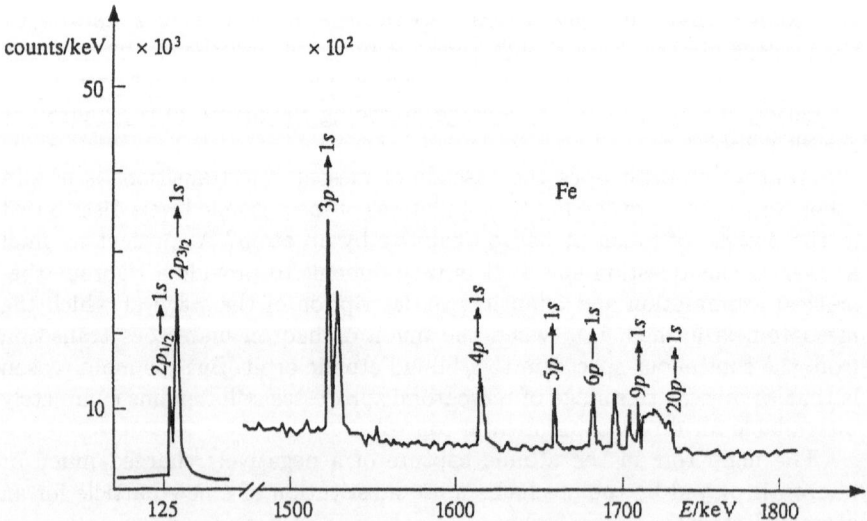

Fig. 13.2. Emission spectrum of mesoatoms of iron [13.1]

principal quantum numbers exceeding 10). At present the situation is different. Figure 13.2 presents part of the X-ray spectrum of muonic atoms of iron, measured at the Swiss meson factory. In all, over a hundred lines corresponding to the K, L, M, N, and O series of transitions were observed. The data shown in the figure reveal that the cascade of transitions starts at very distant orbits ($n > 14$), at least with $n_i = 20$. According to modern ideas, most of the muons and hadrons first land in states with very large principal quantum numbers n, and the distribution over n has a long tail (extending in muonic atoms up to $n \approx 50$).

13.2 Cascade of Electromagnetic Transitions in Mesoatoms: Influence of the Medium on a Mesoatomic Cascade

The main characteristic of a cascade is the distribution of level populations in the mesoatom. The *population of level P_{nl}* is the sum of the number of mesons happening to be in the $|nl\rangle$ state either immediately, when the mesoatom was formed, or owing to electromagnetic transitions from higher orbits. Usually, populations are normalized so that they are reduced to a single act of mesoatom formation (to a single muon or hadron). In introducing the concept of level populations of a mesoatom it is important to stress that since one and the same meson can visit many levels in the course of the cascade, the quantities P_{nl} do not add up to unity:

$$\sum_{nl} P_{nl} > 1 \,. \tag{13.5}$$

In this connection, it is useful to supplement the concept of population with the concept of the fraction of mesons that are absorbed by the nucleus or decay when they occupy level $|nl\rangle$:

$$\omega_{nl} = P_{nl} \frac{\Gamma_{\text{abs}}(nl) + \Gamma_0}{\Gamma_{\text{tot}}(nl)} \,. \tag{13.6}$$

Here

$$\Gamma_{\text{tot}}(nl) = \Gamma_{\text{abs}}(nl) + \Gamma_{\text{el}}(nl) + \Gamma_0 \tag{13.7}$$

is the total width of the $|nl\rangle$ level; $\Gamma_0 = \hbar/\tau$ is the contribution to it due to the spontaneous decay of the meson; $\Gamma_{\text{abs}}(nl)$ and $\Gamma_{\text{el}}(nl)$ are the partial widths of the $|nl\rangle$ level corresponding to absorption of the meson by the nucleus and to elctromagnetic transition from level $|nl\rangle$ to all lower-lying states. Since each meson, in whatever way the cascade chooses to proceed, is ultimately absorbed by the nucleus or decays, the parameters ω_{nl}, unlike the populations P_{nl}, are normalized to unity:

$$\sum_{nl} \omega_{nl} = 1 \,. \tag{13.8}$$

At the same time as using the partial widths, we shall also make use of the concept of the *transition rate*: $\lambda_{el}(nl) = \Gamma_{el}(nl)/\hbar$ and so on. The transition rate has the dimension of inverse time and shows the number of corresponding events per unit time. The total rate of electromagnetic transition from the $|nl\rangle$ state is the sum of the rates of partial transition to the individual lower-lying states:

$$\lambda_{el}(nl) = \sum_{n' < n} \sum_{l'} \lambda_{el}(nl \to n'l'). \tag{13.9}$$

Here each of the terms is the sum of the radiative transition rate λ_γ and the Auger transition rate λ_A:

$$\lambda_{el}(nl \to n'l') = \lambda_\gamma(nl \to n'l') + \lambda_A(nl \to n'l'). \tag{13.10}$$

These two types of transition differ essentially. The electron shell of the mesoatom takes practically no part in the radiative transition of the meson. On the contrary, Auger transition consists precisely in the transfer of released energy from the meson to an atomic electron, which leaves the atom. Although both types of transition are realized in mesoatoms mainly owing to $E1$ transitions, the selection rules for radiative and Auger transitions differ from each other. The following are the most probable transitions:

radiative: $nl \to n' = l; \quad l' = l - 1;$

Auger: $nl \to n' = n - 1; \quad l'l - 1.$ (13.11)

Hence it is seen that radiative transitions in the main favor the population of circular orbits ($l' = n' - 1$).

The rates of radiative and Auger transitions depend in totally different ways on the transition energy and, consequently, also on the principal quantum number of the level from which the transition proceeds (Fig. 13.3). Therefore, at the early stage of the cascade, Auger transitions are dominant, whereas at the concluding stage radiative transitions dominate. From the figure it is seen that transition from one mode to another proceeds earlier, the larger the atomic number of the mesoatom, Z. This is because the rate of Auger transitions weakly depends on Z, while the rate of radiative transitions is approximately proportional to Z^4; therefore the intersection point of the λ_γ and λ_A curves is shifted toward large n as Z increases.

For calculations of the level populations in the cascade process a set of linear algebraic equations is used:

$$P_{nl} = C_{nl} + \sum_{n' > n} \sum_{l'} P_{n'l'} \frac{\Gamma_{el}(n'l' \to nl)}{\Gamma_{tot}(n'l')}, \tag{13.12}$$

where C_{nl} are the initial level populations (relation (13.12) is usually called the *basic cascade equation*). The parameters Γ_{el} and Γ_{tot}, occurring in the right-hand part of the equation, are calculated by making use of some or other, more or less realistic wave functions of the mesoatomic states and (for

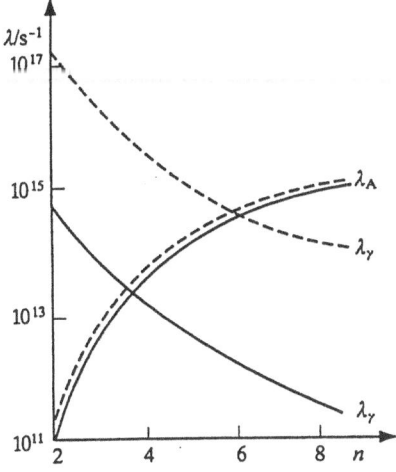

Fig. 13.3. Dependence on n of rates of radiative and Auger transitions of a muon between circular orbits in the mesoatoms of oxygen (*solid lines*) and bromide (*dashed lines*) [13.2]

hadron atoms) of the optical potentials for the interaction between the hadron and the nucleus; in certain cases the total widths of low-lying levels of hadron atoms, $\Gamma_{\text{tot}}(nl)$, may be taken directly from measurements, in accordance with the line widths in the X-ray spectra of these atoms.

Equation (13.12) is for an isolated mesoatom. When applying it to mesoatoms produced in condensed matter or in strongly compressed gases, it is necessary to introduce into this equation corrections taking into account the influence of the environment on the cascade of meson or hadron transitions in the mesoatom. We shall illustrate this influence with two examples.

In atoms with sufficiently large Z the vacancies in the inner shells caused by Auger transitions at the early stage of the mesoatomic cascade may be restored by the mesoatom capturing electrons from the outer shells of atoms of the environment. The rate of these restoration processes determines the character of the transition in the mesoatomic cascade from the mode where Auger transitions are dominant to the mode with dominant emission of X-ray quanta. It turns out to be impossible to control directly the restoration process of the electron shell during the cascade of mesoatomic transitions, although Auger spectroscopy of mesoatoms does promise possibilities. Until now, all the information on this process can be obtained only by deciphering the X-ray spectra of mesoatoms. Often the restoration rate of the electron shell of the mesoatom is included in the equation of the cascade as a separate parameter to be fitted. Let us be more precise: when rearrangement of the electron shell of the mesoatom during the cascade is taken into account, the structure of (13.12) itself also becomes more complex and as well as the level populations of the meson now involves as dynamic variables the populations of various orbits of the electron shell of the mesoatom.

Another example of the influence of the environment on the mesoatomic cascade concerns hydrogen mesoatoms. A peculiarity of such mesoatoms consists in their being electrically neutral. This allows them to move around freely in a medium and, for instance, to approach at a very close distance the nuclei of atoms of the medium, where they are under the influence of the strong electric field created by the charge of the nucleus. We recall that the levels of hydrogen-like systems, including the levels of the muonic hydrogen atom and, also, of the highly excited levels of the hadronic hydrogen atom, are characterized by a degeneracy relative to the orbital momentum l. The strong Coulomb field of the nuclei which the hydrogen mesoatom closely approaches mixes its states in l (so-called *Stark mixing* of states occurs). This means that the muons or hadrons, which were in a circular orbit $(n; l = n-1)$, are now "distributed" over all the orbits with the quantum number n: starting with the circular orbit $(l = n - 1)$ up to the ns orbit distinguished by the high probability for the particle to be at the center of the mesoatom. In the case of hadron hydrogen atoms this is of particular importance since a hadron in the s state, owing to Stark mixing, starts interacting strongly with the proton long before it reaches the lowest-lying orbits.

The consequences of the Stark mixing of the states of hadron hydrogen atoms are very important for investigations in elementary particle physics. Consider, for example, studies with antiprotons of low energies. In liquid hydrogen, where, unlike in the case of gaseous media, the probability of a $\bar{p}p$ antiproton atom, called a *protonium*, interacting with nuclei of the medium (and at the same time the probability of Stark mixing) is very high, annihilation of antiprotons *at rest* proceeds exclusively from s states. This results in a particularly rigorous selection of the final states in the annihilation process. At the same time, in hydrogen gas chambers, protonium annihilates with a noticeable probability from states with $l \neq 0$. This is revealed, for instance, in direct observations of annihilation events from the $2p$ state, which are performed by recording annihilation products in coincidence with characteristic X-ray quanta due to transitions from the higher levels and leading to population of the $2p$ state of protonium.

Retardation effects in cascades of electromagnetic transitions in mesoatoms give rise to great interest. One such effect, which has been known for a long time and studied quite thoroughly, is the delay of the cascade in muonic atoms occurring when the muon is in the metastable $2s$ state of the mesoatom, from which ordinary one-photon radiative transition to the ground $1s$ state is forbidden by the orbital number. Calculations show that although such an atom is isolated, the metastable $2s$ state decays either via the spontaneous decay channel of the muon itself, or (as in the $2s$ state in the ordinary hydrogen atom or in hydrogen-like ions) via the two-photon channel. Under real conditions, the Stark mixing of $2s$ and $2p$ states (and the subsequent fast spontaneous $2p \rightarrow 1s$ transition), as well as the external Auger effect, in which the excitation energy of the muon atom is transferred to electrons of the atoms of the medium, accelerate the transition of the muon from the $2s$ level to the ground state. In a gaseous medium the rate for these processes is proportional to the gas pressure, and this circumstance makes it possible to keep track of how they influence the delay of the cascade.

Another example of cascade retardation concerns hadron atoms. Experiments show that when negative pions and kaons are stopped in liquid helium, a significant number of these particles decay when they are already on mesoatomic orbits. The fraction of such events, reduced to a single stopped meson, is (taking into account results obtained in different laboratories) between 1.0 and 1.2 % for π^- and between 1.9 and 2.6 % for K$^-$. What is surprising in these data? The lifetimes of the pion and kaon are 2.6×10^{-8} s and 1.24×10^{-8} s, respectively. The stopping times of these particles from the initial velocities, corresponding to the usual conditions of the production of extracted pion and kaon beams at accelerators, together with the atomic cascade times are lower by 2–3 orders of magnitude. In liquid helium, where the density is high, Stark mixing reduces this time down to $\sim 10^{-12}$ s. The relation between the times $\tau \sim 10^{-8}$ s and $\tau \simeq 10^{-12}$ s leaves no chance for 1–2 % of the stopped pions or kaons to decay freely, avoiding absorption by the nucleus of the mesoatom. In this connection, a very long time ago, an assumption (the "Condo hypothesis") was made, according to which a certain fraction of pions and kaons going through the cascade of transitions in a mesoatom are detained in long-lived metastable states of a special sort, absorption from which by the nucleus of the mesoatom is difficult and in which most of them decay. According to Condo's hypothesis these should be circular orbits with very large quantum numbers ($n \geq 30$). Auger transition from such orbits is rendered extremely difficult, since the mean binding energy of the electron in a helium mesoatom (~ 25 eV) requires the muon to undergo transition directly to a very low level, which contradicts the selection rules (13.11). The same holds true for radiative transitions: the allowed transitions are transitions between adjacent circular orbits $(n; l = n - 1) \rightarrow (n - 1, l' = l - 1)$. At large n the energy of such transitions and, consequently, the rate of such transitions are very small.

Not long ago, the Condo hypothesis received confirmation in experiments on time scanning of the decay of K mesons stopping in liquid helium (Fig. 13.4). The mean lifetime of stopped K$^-$ mesons turned out to be $\tau(\mathrm{K}^-) = (0.95 \pm 0.03) \times 10^{-8}$ s, i.e., less than the lifetime of free kaons $[\tau_0(\mathrm{K}^-) = 1.24 \pm 0.03 \times 10^{-8}$ s] or of K$^+$ mesons stopped in liquid helium $[\tau(\mathrm{K}^+) = (1.26 \pm 0.02) \times 10^{-8}$ s]. This enhancement of the decay rate of the kaons is interpreted quite naturally within the framework of the Condo hypothesis. The time dependence of the kaon decay from a metastable level is due, on the one hand, to its spontaneous decay and, on the other hand, to absorption of the kaon by the nucleus directly from this level and downward electromagnetic transition (ultimately to levels where the kaon is absorbed by the nucleus):

$$\Gamma_{nl}(\mathrm{tot}) > \Gamma_0(\mathrm{K}^-) = \frac{\hbar}{\tau_0(\mathrm{K}^-)} \tag{13.13}$$

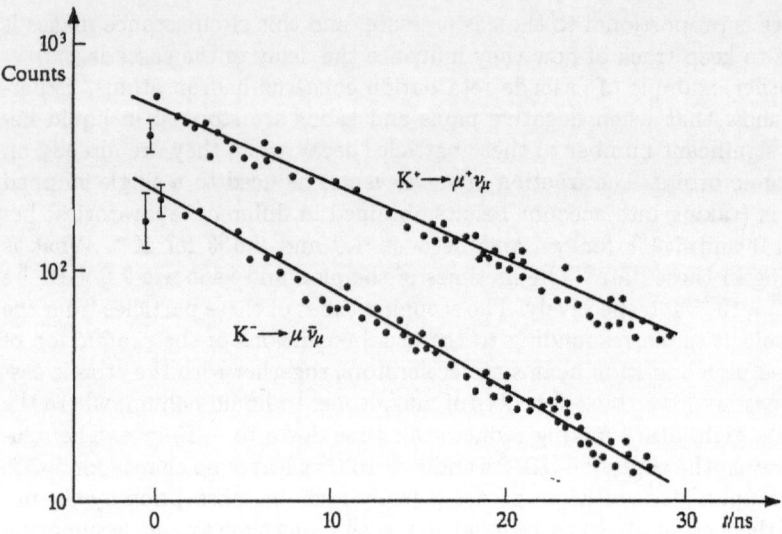

Fig. 13.4. Time dependence of decays of K$^-$ and K$^+$ mesons stopping in liquid helium [13.3]

The existence of a cascade of transitions between the primary metastable level and levels from which the kaon is absorbed by the nucleus results, strictly speaking, in the decay of the delayed kaons being nonexponential. Therefore, relation (13.13) and the equivalent relation for the times,

$$\tau_{nl}(K^-) = \frac{\hbar}{\Gamma_{nl}(\text{tot})} < \tau_0(K^-)\,, \tag{13.14}$$

can be useful only for orientation.

The cascade retardation effect of transitions between high-lying circular levels is especially interesting for studies with antiprotons. The antiproton is stable, and the procedure of time scanning (such as presented in Fig. 13.4), when applied to the yield of antiproton annihilation products in antiproton atoms not only gives the lifetime of long-living metastable states in these atoms, but also permits the identification of the process itself of annihilation of the delayed antiprotons. From the point of view of nuclear physics, the peculiarity of this process consists in that, here, the orbital momentum of the $\overline{N}N$ pair is very high at the moment of annihilation. This is very important, since at present practically no direct information is available on the properties of the annihilation products of the $\overline{N}N$ pair from states with high orbital momenta.

13.3 Methods Involving Pions and Muons in Chemical Studies

In this section we shall examine several examples of the application of pionic and muonic methods in chemical studies to show their relationship with problems of the physics of mesoatoms and of mesoatomic processes.

13.3.1 Analysis of the Composition and Structure of Matter Based on Mesonic X-Rays

Measurement of the positions and intensities of separate lines in the spectra of mesonic X-rays permits us to determine the chemical elements, present in the substance under investigation, to which these lines or, moreover, whole groups of lines (series) belong. This method can be considered a variety of the conventional X-ray method of elemental and structural analysis. However, a number of essential features make the mesonic X-ray method a significant supplement to the usual methods. The special possibilities presented by the mesonic X-ray method arise because the probability of atomic capture by atoms of different elements at the formation stage of mesoatoms exhibits a distribution depending not only on the relative concentrations of the elements in the substance under investigation and the atomic number Z_i of each element, but also on the existence and kind of chemical bonds between the atoms of these elements. Ultimately, this dependence is manifested in the relation between the intensities of individual lines and series in the measured mesonic X-ray spectra. The essential possibilities of applying the mesonic X-ray method for revealing the character of chemical bonds in the samples studied make the construction of a physical theory of atomic capture of muons and hadrons in mixtures and chemical compounds very important. But there still exist many open questions.

The relation between the probabilities of atomic capture of a muon or of a hadron, $W(Z_i)$, by atoms of different chemical elements is determined first by their concentrations c_i in the substance under investigation. It is convenient to exclude this trivial factor from $W(Z_i)$ by using the reduced probabilities of atomic capture, $w(Z_i)$, which are calculated for the same number of nuclei of different elements:

$$w(Z_i) = \frac{W(Z_i)}{c_i} . \tag{13.15}$$

The so-called Z law is known and often applied for orientation in estimating the relationship between the reduced probabilities $w(Z_i)$ for different elements in the mixture or chemical compound. According to this law, the reduced probability of a muon or a hadron being captured by an atom is proportional to the number of electrons in the atomic shell, i.e., to its atomic number Z:

$$\frac{w(Z_1)}{Z_1} = \frac{w(Z_2)}{Z_2}.$$

(13.16)

Actually, significant deviations from this law are observed.

13.3.2 Charge Exchange of π Mesons on Hydrogen Nuclei

One of the most important problems of modern chemistry is that of the hydrogen content in condensed media. Mesochemistry proposes a method exhibiting great selectivity in analyzing the concentration and chemical bonds of hydrogen in various substances. This method makes use of "stopped" π^- mesons, i.e., of π^- mesons that have undergone slowing-down in a substance and have been captured onto Coulomb orbits of mesoatoms. The concluding stage of a cascade of electromagnetic transitions in a pionic hydrogen atom differs essentially from that in the pionic atoms of all other elements. Here, there is no usual absorption of the pion by a nucleus, which proceeds via the strong interaction channel, and therefore all other channels of the pion-nucleus interaction, which usually play no significant part in the ultimate fate of pionic atoms, now become the main processes. Such are the reactions of radiative pion capture by a proton and of pion charge exchange:

$$\pi^- + p \rightarrow n + \gamma,$$

(13.17)

$$\pi^- + p \rightarrow \pi^0 + n.$$

(13.18)

The neutral π^0 meson produced in the charge-exchange reaction decays via the channel $\pi^0 \rightarrow 2\gamma$, so by measuring the ratio of the yield of characterstic pairs of hard γ quanta ($E_\gamma \approx 70\,\mathrm{MeV}$) and the yield of single γ quanta in reaction (13.17) ($E_\gamma > 100\,\mathrm{MeV}$), it is possible to find the ratio between the probabilities of reactions (13.18) and (13.17), which is the so-called *Panofsky ratio* P_H:

$$P_H = \left.\frac{Y(\pi^-, \pi^0)}{Y(\pi^-, \gamma)}\right|_{\mathrm{exp}} = 1.53 \pm 0.02.$$

(13.19)

As we see, over 60 % of the pions captured from Coulomb orbits of the hydrogen mesoatom finish their lives with charge exchange. Only one more nucleus – ^{3}He – is known for which the output of the charge-exchange reaction of stopping π mesons is relatively high (~ 15–$17\,\%$); in all other nuclei the charge-exchange yield does not exceed 0.5 %. The charge exchange of stopped pions in various hydrogen-containing substances turns out to be an extremely sensitive means not only for establishing the hydrogen content in them, but also for clarifying whether the hydrogen in these substances is free or whether it is chemically bound.

13.3.3 Polarization and Depolarization of Muons in Muonic Atoms

In laboratory investigations the muon sources are pions decaying via weak interaction channels:

$$\pi^- \to \mu^- + \bar{\nu}_\mu; \quad \pi^+ \to \mu^+ + \nu_\mu. \tag{13.20}$$

In accordance with the physical laws of weak interactions, the neutrino and antineutrino are always totally polarized: the neutrino ν_μ is left-polarized (its helicity, i. e., the projection of its spin onto its momentum p, is $\langle s_\nu \rangle_p = -1/2$); the antineutrino $\bar{\nu}_\mu$ is right-polarized ($\langle s\bar{\nu} \rangle_p = +1/2$). The pion is a spinless particle. Therefore, in decays (13.20) the μ^- and μ^+ muons produced are also (in the rest frame of the pion) totally polarized. Recalculating to the laboratory reference system and taking into account the loss of polarization due to the collimation of muon beams reveals that at pion energies and for geometrical conditions typical of meson factories and other laboratories of similar profiles, polarization of the muons produced is not total, but the deviation does not exceed 20 %. In the process of stopping in matter, further depolarization of the muons is not significant. Thus, negative muons are captured onto mesoatomic orbits with a high degree of longitudinal polarization – of over 80 %.

What happens with the spin of a muon as the cascade of its mesoatomic transitions develops? At the early stage of the cascade Auger transitions dominate, since they proceed rapidly as compared with the precession time of the muon spin in the magnetic field of the nucleus. In other words, the widths of the upper levels of muonic atoms, to which Auger transitions give a dominant contribution, are large in comparison with the difference between the levels of the fine structure of the muonic atom. In such conditions the initial polarization of the muon is conserved. The picture changes qualitatively when the cascade approaches the levels at which Auger transitions give up their dominant role to radiative transitions. Calculations show that the radiative widths of the levels of a muonic atom are always smaller than their fine splitting. Under such conditions, strong depolarization of the muon takes place. If the spin of the nucleus of a mesoatom is zero, then calculations reveal that the muon retains only 16–17 % of its initial polarization upon reaching its K orbit. If the spin of the nucleus differs from zero, then owing to the hyperfine interaction it retains even less. The residual muon polarization may be determined from the forward–backward asymmetry of the angular distribution of electrons produced in muon decays: $\mu^- \to e^- + \nu_\mu + \bar{\nu}_e$. Utilization of this phenomenon in mesochemistry is based on the fact that in condensed media and compressed gases the interaction of a muonic atom with the environment may influence (via the electron shell of the muonic atom or directly via the interaction of the muon magnetic moment with the internal magnetic fields in the sample) the residual polarization of the muon.

We recall that the free muon lives for only $2\,\mu s$. With muon capture by the nucleus taken into account, the mean lifetime of a muonic atom on the K orbit is even shorter, and it is shorter the more the wave function of the muon on the K orbit overlaps with the internal region of the nucleus, i.e., the heavier the nucleus. For example, in nickel $\tau_\mu(_{28}\mathrm{Ni}) = (0.162\pm0.001) \times 10^{-6}$ s, in barium $\tau_\mu(_{56}\mathrm{Ba}) = (0.0944 \pm 0.0007) \times 10^{-6}$ s, in uranium $\tau_\mu(^{235}_{92}\mathrm{U}) = (0.0665 \pm 0.0042) \times 10^{-6}$ s. During this short time interval the muonic atom of the Zth element behaves in a medium like an analog of the atom of the $(Z - 1)$st element, owing to the very small dimensions of its K orbit; for instance, it takes part in chemical reactions like an ordinary atom of this $(Z - 1)$st element. At the same time it is a "tagged" atom, which reveals itself by the muon decay. Owing to its residual polarization the muonic atom remembers the initial direction of its muon spin. If a magnetic field H, perpendicular to this direction, is applied to the sample, the magnetic moment of the muonic atom starts precessing, and, consequently, the whole diagram of the angular distribution of electrons from muon decays rotates about the magnetic field vector. If the spin of the nucleus and the electron shell is zero (a diamagnetic atom), then the angular velocity of such a rotation is simply the Larmor rotation frequency of the muon spin:

$$\omega_\mu = \frac{eH_\perp}{2m_\mu c}. \tag{13.21}$$

An electron detector equipped with fast electronics and detecting outgoing particles within a certain solid angle records not the exponential muon decay $N_e(t) \sim \exp(-t/\tau_\mu)$, but the decay modulated by precession of its magnetic moment:

$$N_e(t) = N_e(0)e^{-t/\tau\mu}\left[1 + a\cos(\omega_\mu t + \varphi)\right], \tag{13.22}$$

where the coefficient a is related to the residual polarization of the muon on the K orbit. If, on the other hand, chemical processes involving a tagged μ^- atom at a certain moment make it paramagnetic, then the precession frequency will be determined by the total magnetic moment of the atom, and this immediately affects the time dependence of the detection of electrons from muon decays.

13.3.4 Muonium

Hitherto we have dealt with mesoatomic processes involving negatively charged particles. Relevant to these are investigations of muonium and of processes involving its participation. Muonium, μ^+e^-, is a system consisting of a positively charged muon and an electron. The exponential decay of the muon, $\mu^+ \rightarrow e^+ + \nu_e + \bar{\nu}_\mu$, if observed by the output of positrons in a magnetic field perpendicular to the muon spin, is modulated by periodic oscillations such as (13.22) owing to precession of the magnetic moment of the muonium. If during the lifetime of the μ^+ ($\sim 2.2 \times 10^{-6}$ s) the free muonium takes part in a chemical reaction, this is immediately seen from the character of the precession. Thus, it is possible to measure the rate of chemical reactions with participation of the muonium and, consequently, to judge about the rate of chemical reactions with the participation of hydrogen, of which muonium is a direct analog.

Fourteenth Lecture

14.1 The Formation of Mesomolecules

The simplest molecular system, the molecular hydrogen ion ppe $\equiv$ H_2^+, was mentioned in the Introduction. The mesomolecular hydrogen ion (mesoion) ppμ $\equiv$ μH_2^+, where a negatively charged muon μ^- is substituted for the electron, is its direct analog. The existence of the μH_2^+ mesoion as a stable system was first contemplated in the early years of investigations with mesons. The idea was also then put forward of the muon catalysis of nuclear fusion (F. Frank, 1947). The μH_2^+ mesoion is a very compact system. While the equilibrium distance between the nuclei in an ordinary molecular hydrogen ion H_2^+ is of the order of 10^{-8} cm, in the μH_2^+ mesoion it is two orders of magnitude smaller, which corresponds to the ratio between the muon and electron masses, $m_\mu/m_e = 207$. Calculations show that when the hydrogen nuclei composing the μH_2^+ mesoion approach each other at this distance, they have time, during the lifetime of the muon, to overcome the Coulomb barrier between them and to take part in nuclear interactions.

In the next section we shall separately consider the situation with the problem of μ catalysis and the relation between the problem of catalysis and the physics of charged particle interaction with matter. Here, also, we shall examine the main properties of hydrogen mesomolecules.

We shall turn to formula (C.24) for the effective potential energy of the interaction between the hydrogen nuclei in the H_2^+ molecular ion. Let X be any one of the negatively charged particles e, μ, or π. We shall consider the mass of particle x to be much smaller than the proton mass: $m_X \ll m_p$. Then, formula (C.24) will be rewritten without alteration for the mesomolecular ion XH_2^+, also:

$$U_{\text{eff}}(R) = \frac{e^2}{R} + \varepsilon(R)\,. \tag{14.1}$$

The second term in this expression, with account taken of the scaling effect on the wave functions of hydrogen-like systems, will be written in the form

$$\varepsilon(R) = \frac{e^2}{a_0}\frac{m_\nu}{m_e}f(\xi)\,, \tag{14.2}$$

where we have introduced the dimensionless variable

$$\xi = R/a_X; \tag{14.3}$$

here

$$a_X = \frac{m_e}{m_X} a_0 \tag{14.4}$$

is the atomic unit of length reduced by a factor of m_X/m_e, and $f(\xi)$ is a common function of ξ for all H_X systems. Now, the whole expression (14.1) may be rewritten making use of this dimensionless variable:

$$U_{\text{eff}}(R) = \frac{e^2}{a_0} \frac{m_X}{m_e} \left[\frac{1}{\xi} + f(\xi) \right]. \tag{14.5}$$

The minimum of the function $1/\xi + f(\xi)$ (let it be at point ξ_0) determines the equilibrium distance between the nuclei, R_0:

$$(R_0)_X = \frac{1}{m_X/m_e}(R_0)_e, \qquad (R_0)_e = a_0\xi_0; \tag{14.6}$$

its second derivative at point ξ_0 is the rigidity coefficient of the molecular ion relative to oscillations of the nuclei in the lowest state of their rotational motion $(J = 0)$:

$$k = [1/\xi + f(\xi)]''_{\xi_0} \left(\frac{m_X}{m_e} \right)^3 \frac{e^2}{a_0^3}. \tag{14.7}$$

The oscillation frequency of the nuclei is related to the rigidity coefficient by the formula

$$\omega = \sqrt{\frac{k}{\mu}}, \tag{14.8}$$

where $\mu = M_1 M_2/(M_1 + M_2)$ is the reduced mass of the molecular ion, and M_1 and M_2 are the masses of the nuclei of the hydrogen isotopes composing the ion considered. Hence the scaling relation for the frequencies:

$$\omega_X = \left(\frac{m_X}{m_e} \right)^{3/2} \omega_e. \tag{14.9}$$

The oscillation frequency of the nuclei in the XH_2^+ mesoion is much greater, than in the ordinary molecular hydrogen ion.

The relations presented above between the linear and energy parameters of the ordinary and of the mesic molecular ions are really estimations. First, even in μH_2^+, the lightest of the mesomolecular ions, where the negative muon assumes the role of the electron, the ratio of the muon mass to the proton mass is not such a small quantity $(m_\mu/M_p \approx 1/7)$, and therefore the corrections to the adiabatic approximation (within the framework of which the initial ratio (C.24) was obtained) turn out to be significant. Second, in our estimations we have disregarded rotations of the molecular ion. But in transition from the ordinary molecular hydrogen ion to the mesomolecular ion XH_2^+ the change in the spectrum of its rotational levels is not the same as in the case of nuclear vibration frequencies:

$$\frac{\hbar^2 J(J+1)}{2\mu R_0^2} \sim \left(\frac{m_X}{m_e}\right)^2, \tag{14.10}$$

i. e., the scaling coefficient is not the same over the whole spectrum of rotational and oscillation ("ro-vibrational") excitations of mesomolecular ions.

Table 14.1 presents parameters of molecular ions of hydrogen isotopes, calculated by various authors and by various methods; the most detailed data are available for the $dd\mu$ and $dt\mu$ ions, which is related to their special role in investigations relevant to the μ catalysis of nuclear fusion. We note that the conditions of the mutual disposition and motion of the nuclei in the $pp\mu$, $pd\mu$, $dd\mu$, etc. mesoions are very similar to those known in astrophysics for white dwarfs and in nuclear physics for hot plasma in thermonuclear installations with magnetic or inertial confinement. (Tokamak, laser heating of plasma and others). Indeed, a density $\rho \sim 10^{-30}\, 1/cm^3$, i. e., 10^7 LHD (LHD $= 4.25 \times 10^{22}$ atoms/cm^3 is the density of liquid hydrogen) corresponds to the mean distance between nuclei of the order of 10^{-10} cm. On the other hand, 100 eV, which is of the same order of magnitude as the energy of nuclear oscillations in the μ mesomolecular hydrogen ion, amounts to a temperature of the order of a million degrees.

Table 14.1. Dissociation energy of μ molecular ions of hydrogen isotopes (in eV) in various states (J, v) of rotational and oscillation excitation

J, v	ppμ	ddμ	dtμ	ttμ
0, 0 (ground state)	253	325	319	362
0, 1	–	36	35	84
1, 0	107	227	232	289
1, 1	–	1.97	0.66	45.2

The first observation of the formation of mesomolecular compounds took place in 1957 in an experiment performed by the group of Alvarez. In a liquid-hydrogen bubble chamber containing an admixture of deuterium, toward which negatively charged pions and muons were directed, events were observed in which muons of a characteristic energy equal to 5.3 MeV were produced at the stopping point of the primary particle. They could not be produced in the decays of stopping pions, $\pi^- \rightarrow \mu^- \bar{\nu}_\mu$ (here, the energy of the muon produced would be only 4.1 MeV). It became clear that the

5.3 MeV muons could only be due to the reaction of nuclear fusion following the formation of the pdμ mesomolecule:

$$\mathrm{pd}\mu \to {}^3\mathrm{He} + \mu^- \,. \tag{14.11}$$

Comparison of this reaction with the known fusion reaction studied at accelerators under usual conditions,

$$\mathrm{p} + \mathrm{d} \to {}^3\mathrm{He} + \gamma \; (E_\gamma = 5.5\,\mathrm{MeV}) \,, \tag{14.12}$$

allows one to conclude that reaction (14.11) proceeds as if "internal conversion" of the energy released in nuclear fusion occurs, and the muon, which at the time of fusion takes place on a mesomolecular orbit of the $\mu \mathrm{H}_2^+$ ion, carries away nearly all this energy (with the exclusion of a small fraction taken up by the ${}^3\mathrm{He}$ recoil nucleus).

The Alvarez experiment proved to be very important for subsequent mesomolecular investigations. This experiment demonstrated that: (1) mesomolecules actually exist; (2) upon forming a mesomolecule, negatively charged muons are then freed, and having undergone stopping they may again lead to the formation of a mesomolecule, i. e., it was shown experimentally that the muon is capable of serving as a catalyzer; (3) the detection of characteristic products of nuclear fusion serves as the method for finding mesomolecules. In subsequent years a definitive contribution to experimental and theoretical investigation of physical processes involving the participation of mesomolecules was provided by physicists of the Joint Institute for Nuclear Research at Dubna. Especially interesting results were obtained in studies of dd fusion:

$$\mathrm{dd}\mu \begin{array}{l} \longrightarrow \mathrm{p} + {}^3\mathrm{H} + \mu^- \\ \longrightarrow \mathrm{n} + {}^3\mathrm{He} + \mu^- \,. \end{array} \tag{14.13}$$

It turned out that the rate of this reaction depends on the temperature of the medium (Fig. 14.1). At the beginning, such a result seemed strange. Now, it is already generally recognized that the discovery of the temperature dependence of the dd fusion rate indicated the way to drastic enhancement of the efficiency of the muonic catalysis and marked the beginning of broad systematic studies in this field.

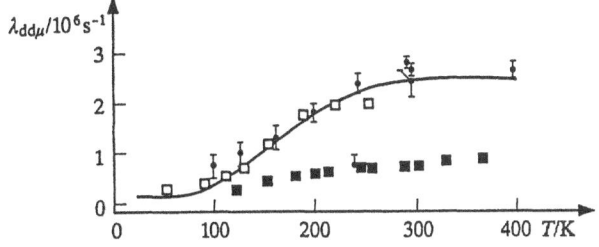

Fig. 14.1. Formation rate of the ddμ muonic molecule versus the temperature of the medium [14.1]. The data following the solid curve were obtained later in other laboratories

14.2 Muonic Catalysis of Nuclear Fusion

The maximum number of cycles per muon that has been obtained in experiments on μ catalysis (St. John, Los-Alamos, 1986), amounts to about 150. This is achieved in deuterium–tritium mixtures, where dt fusion takes place:

$$\mathrm{dt}\mu \rightarrow \alpha + \mathrm{n} + \mu^- \quad (Q = 17.6\,\mathrm{MeV}). \tag{14.14}$$

The particularly high energy released in each cycle of this process and the large number of cycles per muon represent a lucky combination of the features of dtμ fusion, which leave it without any competitors among all the other reactions of μ catalysis, from the point of view of future utilization in power engineering.

Figure 14.2 presents the μ-catalysed fusion chain in a mixture of deuterium and tritium. The catalysis efficiency depends on the time length of the cycle, τ_c, since precisely this quantity, when compared with the lifetime of the muon itself, τ_0 (or, which is the same, its inverse – the mean cycling rate $\lambda_c = 1/\tau_c$, as compared with the mean spontaneous decay rate of the muon, $\lambda_0 = 1/\tau_0 = 0.45 \times 10^{-6}\,\mathrm{s}^{-1}$) determines the mean number of cycles per muon. What forms the cycle time? Let us follow the muon with the aid of the chain depicted in Fig. 14.2. We shall take into account the rates of all pair-collision processes being proportional to the density of the medium; the numerical values for the rates of such processes indicated in the picture are related to a medium with a density equal to the density of liquid hydrogen.

When a muon lands in a substance, it is decelerated according to the general laws of ionization stopping, and then it interacts with $\mathrm{D_2}$, $\mathrm{T_2}$, and DT molecules producing muon atoms of the hydrogen isotopes, dμ and tμ. This stage takes about 10^{-12} s. In the figure no processes related to the

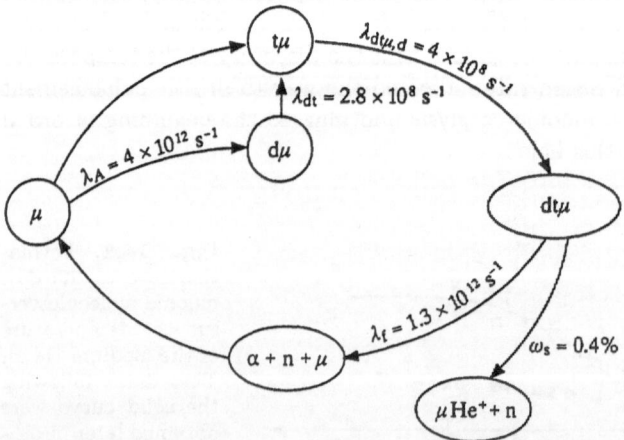

Fig. 14.2. The μ catalysis cycle in a mixture of deuterium and tritium; the rates of collisional processes are given for the density of liquid hydrogen

cascade of electromagnetic transitions of the muon in the $d\mu$ and $t\mu$ atoms are reflected. These processes are also very rapid (see Lecture 13); they are completed within a time of about 10^{-11} s. The $d\mu$ and $t\mu$ atoms produced, which are compact neutral objects, start to migrate through the molecules of the medium. Upon approaching a tritium nucleus in a DT or T_2 molecule, the $d\mu$ atom readily gives up its muon, since, owing to the isotopic effect, the energy binding the muon to the tritium is stronger than that binding it to the lighter deuteron:

$$d\mu + t \rightarrow t\mu + d. \tag{14.15}$$

Thus, before the formation of a mesomolecule, the muon is most often in a $t\mu$ atom. This $t\mu$ atom interacts with the deuteron present in a D_2 or DT molecule and forms a $dt\mu$ mesomolecular ion, or a $tt\mu$ mesomolecular ion together with the triton of a DT or T_2 molecule. However, it turns out that the formation rate of the first of these ions is two orders of magnitude greater than that of the other (the formation rate $\lambda_{tt\mu}$, which is not presented in Fig. 14.2, amounts to approximately 3×10^6 s^{-1}). Moreover, the formation rate of the $dd\mu$ mesomolecule (produced in a $d\mu + D_2$ or $d\mu + DT$ collision) is also lower than $\lambda_{dt\mu}$ by two orders of magnitude. Why, then, is the formation of $dt\mu$ mesomolecules so strongly singled out among other similar molecule production processes?

The reason for this consists in the resonance formation mechanism of muonic molecules of the hydrogen isotopes. It was proposed by Vesman (Dubna, 1967) as a method for solving the already familiar problem of the temperature dependence of $dd\mu$ fusion. Starting from this idea, Gershtein and Ponomarev 10 years later showed that the resonance mechanism should be especially strong in the formation of the $dt\mu$ molecule; at the same time the number of 100 was first mentioned as the number of cycles per muon. Theoretical calculations of the mesomolecular formation process in mixtures of hydrogen isotopes performed by several groups in different countries reveal the details of how the resonance mechanism of $dt\mu$ fusion is realized.

A $t\mu$ mesoatom interacting with a D_2 molecule does not simply take a deuteron away from the molecule,

$$t\mu + D_2 \rightarrow dt\mu + d + 2e, \tag{14.16}$$

leaving the other deuteron free. Such a process (it is called direct meso-molecule formation) is possible, but it is not the main one. In the main, a molecular ion $dt\mu$ becomes, together with this second deuteron, a composite part of the newly produced large $(dt\mu)dee$ complex, which is an analog of the ordinary hydrogen molecule. The dimensions of the $dt\mu$ ion in this large two-electron molecule are of the order of 5×10^{-11} cm, the dimensions of the whole complex are similar to those of an ordinary molecule, i.e., of the order of 10^{-9} cm (Fig. 14.3). The $dt\mu$ ion in this large molecule serves as a heavy Coulomb center, something like a superheavy hydrogen nucleus, with a charge $+e$ and a mass over five times greater than the proton mass.

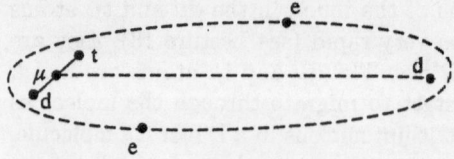

Fig. 14.3. Diagram of the structure of the $(dt\mu)$dee mesomolecule

Thus, the mesomolecular complex $(dt\mu)$dee has both mesomolecular and ordinary molecular degrees of freedom. Excitations associated with different degrees of freedom combine with each other to create the peculiar pattern of the whole excitation spectrum of this complex (Fig. 14.4). We shall pay special attention to the $\varepsilon(J = 1, v = 1) = \varepsilon_{11}$ level situated in the immediate vicinity of the dissociation threshold of the molecular ion: $\varepsilon_{11} = 0.66\,\text{eV}$ (see Table 14.1). In the right-hand part of Fig. 14.4, the spectrum is shown of ro-vibrational excitations of the $(dt\mu)$dee molecule, which are related to the motion of the deuteron and the $dt\mu$ as a whole. When these excitations are superimposed on to the internal excitation ε_{11} of the $dt\mu$ molecular ion itself, certain levels of the entire $(dt\mu)$dee complex are above the dissociation threshold of the $dt\mu$ molecular ion. To them correspond disintegrating states of the large molecule,

$$\left[(dt\mu)^*\text{dee}\right]^* \to t\mu + \text{D}_2 . \tag{14.17}$$

We are more interested, however, in reading this reaction in the opposite direction:

$$t\mu + \text{D}_2 \to \left[(dt\mu)^*\text{dee}^*\right] , \tag{14.18}$$

in which the free $t\mu$ mesoatom interacts with the D_2 molecule. At certain energies of their relative motion there appears, as in the case of resonance nuclear reactions, a compound (quasistable) $[(dt\mu)^*\text{dee}]^*$ system. This is the resonance mechanism of mesomolecule formation. Like the mechanism in (14.18), it is also responsible for the formation of the mesonic molecules $dd\mu$:

$$d\mu + \text{D}_2 \to \left[(dd\mu)^*\text{dee}\right]^* . \tag{14.19}$$

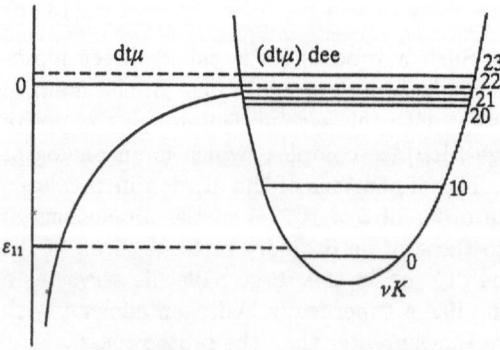

Fig. 14.4. Diagram of the excitation of various degrees of freedom for the $(dt\mu)$dee mesomolecular complex

The widths of these resonances are very small, but the resonance cross sections of molecule formation are significantly larger than the cross sections of direct processes such as (14.10).

Whether the resonance mechanism plays a part in one or another reaction or not depends on whether the resonance level in the composite system produced is just in the interval of excitation energies corresponding to the kinetic energy of a free $d\mu$ or $t\mu$ mesoatom at the moment it interacts with a molecule of the medium. Nature has decided to create such a level in the mesomolecular complexes $(dt\mu)$dee and $(dd\mu)$dee; this is due to the existence of a weakly bound excited state $(J = 1, v = 1)$ in the $dt\mu$ and $dd\mu$ mesoatoms. Truly (see Table 14.1), in the $dd\mu$ ion it lies much deeper, and, consequently, states of higher quantum numbers land in the resonance energy region of the $(dd\mu)$dee molecule, instead of the $(dt\mu)$dee molecule. This means that the formation of such states in $d\mu + D_2$ collisions requires a more serious restructuring of the colliding objects, so ultimately the formation rate of $dd\mu$ mesomolecules is significantly inferior to the particularly high value of $\lambda_{dt\mu}$. Let us make a comparison: $\lambda_{dt\mu,d} = 4 \times 10^8\,\text{s}^{-1}$, $\lambda_{dd\mu,d}(30\,\text{K}) = 5 \times 10^4\,\text{s}^{-1}$, $\lambda_{dd\mu,d}(300\,\text{K}) = 3 \times 10^6\,\text{s}^{-1}$.

Thus, the idea of a resonance mechanism provides answers to two fundamental questions in the physics of the muon-catalyzed fusion concerning: the reason for the extremely high rate of $dt\mu$ mesomolecule formation in D–T mixtures, and the nature of the temperature dependence of the mesomolecule formation probability. Think of the figures: one Kelvin is approximately $10^{-4}\,\text{eV}$. This means that the entire $T = 25\text{--}400\,\text{K}$ range of temperatures shown in Fig. 14.1 occupies only 0.05 eV. Hence we see the high precision necessary in the position of the resonance level in order to predict theoretically the temperature dependence of catalysis. Modern calculations of energy spectra and of the wave functions of mesomolecules are performed by taking into account, in addition to the Coulomb interaction, that nuclei are not point-like, hyperfine interactions, the polarization of vacuum, and other effects. The most accurate of these calculations approach the level of experimental requirements. Thus, the postition of the $(J = 1, v = 1)$ level in the $dd\mu$ mesoion, derived from experimental data on the temperature dependence of the parameter $\lambda_{dd\mu}(T)$, is $\varepsilon_{11} = 1.9659 \pm 0.0003\,\text{eV}$; the most accurate theoretical values achieved in calculations performed by various methods amount to 1.9654 and 1.9668 eV.

Let us return to Fig. 14.2. The figure does not reflect transitions of the $dt\mu$ mesoion from the $(J = 1, v = 1)$ state, in which it was produced, to lower-lying states. Such transitions are very rapid and take about $10^{-12}\,\text{s}$ proceeding via single-step or multi-step (with the ejection of several electrons) Auger transitions. The nuclear fusion reaction $dt\mu \rightarrow \alpha + n + \mu$ occurs when the mesoion is either in the excited $(J = 0, v = 1)$ state, or in the ground state $(J = 0, v = 0)$. This is also a very rapid process: thus, the width Γ_j of the $(J = 0, v = 0)$ level, related to nuclear fusion, amounts to about $10^{-3}\,\text{eV}$

(i. e., the rate of the nuclear reaction from this mesoion level $\lambda_f \approx 10^{12}\,\mathrm{s}^{-1}$). The rate of fusion taking place directly from the initial ($J = 1, v = 1$) state is sigificantly lower, since, here, the centrifugal barrier $\hbar^2 J(J + 1)/2\mu R^2$ is added to the Coulomb barrier between the d and t nuclei, the slow nuclear fusion process from the ($J = 1, v = 1$) state cannot compete with downward Auger transitions from this state.

Thus, upon passing through a chain of various processes terminating with a nuclear reaction, the muon is again free. The μ-catalysis chain is completed. Further, the muon is again slowed down, it undergoes capture by an atom, and so on. We see that the time required for the whole cycle, τ_c, is determined by the rates of two processes: transfer of the muon from the dμ to the tμ atom, and formation of the dtμ mesomolecule. All the other processes are several orders of magnitude more rapid and give practically no contribution to t_c. Therefore, the mean time of the dtμ cycle in a D–T mixture can be written in the form

$$\tau_c = \frac{1}{\lambda_c} \approx \frac{C_d}{\lambda_{dt} C_t} + \frac{1}{\lambda_{dt\mu} C_d}, \tag{14.20}$$

where C_d and C_t are the relative concentrations of the deuterium and tritium nuclei ($C_d + C_t = 1$). When the concentrations C_d and C_t are comparable, τ_c is of the order of 10^{-8} s. This time seems sufficient for the muon to complete 200 catalysis cycles, and for the energy of nuclear fusion released to exceed 3 GeV per muon. The nuisance, however, is that we have forgotten an inconspicuous, at first sight, effect, the so-called *muon sticking*.

In the nuclear fusion process considered, the main three-particle channel $\alpha + n + \mu^-$ is realized together with the two-particle channel $\mu\mathrm{He}^+ + n$, although the latter has a significantly lower probability:

$$\mathrm{dt}\mu \quad \begin{array}{l} \raise2pt\hbox{$\rightarrow \alpha + n + \mu^-$} \\ \lower2pt\hbox{$\hookrightarrow \mu\mathrm{He}^+ + n\,.$} \end{array} \tag{14.21}$$

Here, the muon "sticks" to the α particle and produces together with it a muonic helium ion $\mu\mathrm{He}^+$, which is a typical hydrogen-like system. In the two-particle channel, the fusion energy ($Q = 17.6\,\mathrm{MeV}$) is shared between the $\mu\mathrm{He}^+$ ion and the neutron, inversely proportionally to their masses, so the initial kinetic energy of the $\mu\mathrm{He}^+$ ion is about 3.5 MeV. With the sticking effect taken into account, the mean number of catalysis cycles, Y, is no longer simply the ratio of the cycling rate λ_c to the spontaneous muon decay rate λ_0, but is a somewhat more complex expression:

$$Y = \frac{\lambda_c}{\lambda_0 + \lambda_c \omega_s}, \tag{14.22}$$

which also involves the sticking coefficient ω_s, the relative probability of the reaction dt$\mu \rightarrow \mu\mathrm{He}^+ + n$ occuring in a dtμ fusion event. Taking into account the actual relation between the rates λ_c and λ_0, it can be seen that even in the case of a very small coefficient, $\omega_s = 0.5\,\%$, the sticking effect reduces the catalysis efficiency by a factor of two. For this reason, experimental and theoretical studies of the sticking effect and, also, searches for and studies of mechanisms capable of somehow weakening this effect have determined during recent years one of the main lines of activity in the whole problem of the muon-catalyzed fusion.

The sticking effect is weakened by the possibility of the fast $\mu\mathrm{He}^+$ ion losing, during its slowing down, the muon in collisions with nuclei of the molecules of the medium. This muon continues playing its part of a catalyzer of nuclear fusion. To calculate theoretically the probability of such reactivation of the muon it is necessary to have at one's disposal a great number of data on the cross sections of various elementary processes involving the $\mu\mathrm{He}^+$ ion. To this end both direct (even though cumbersome) quantum-mechanical calculations and indirect methods are applied. For example, if stripping of the muon proceeds from the main state of the $\mu\mathrm{He}^+$ ion, then the cross section of this process,

$$\mu\mathrm{He}^+ + \mathrm{D(T)} \rightarrow \alpha + \mathrm{D(T)} + \mu^-\,, \tag{14.23}$$

can be found by the scaling method, making use of direct experimental data of the cross section of a similar process with the ordinary hydrogen ion:

$$\mathrm{He}^+ + \mathrm{H} \rightarrow \mathrm{He}^{++} + \mathrm{H} + \mathrm{e}\,. \tag{14.24}$$

However, reactivation of the muon is actually a complex cascade process involving, as well as muon stripping, excitation and de-excitation of the $\mu\mathrm{He}^+$ ion in collisions, Stark mixing of its states in the nuclear Coulomb field of the medium, spontaneous radiative transitions from its excited levels, and the external (with transfer of the excitation energy to an electron of the medium) Auger effect, and all this at the same time as the usual ionization slowing down of the $\mu\mathrm{He}^+$ ion, which is a heavy particle. A special place among such processes is occupied by muon stripping from the excited orbits of the $\mu\mathrm{He}^+$ ion, since the cross section of such stripping (like the geometric cross section of the ion itself) rises rapidly as the principal quantum number of the orbit increases; in this case the actual excitation of the $\mu\mathrm{He}^+$ ion is induced during its collisions preceding the stripping. Calculations show that at the density of liquid hydrogen the slowing down of the $\mu\mathrm{He}^+$ ion from the initial velocity down to velocities at which stripping is no longer possible just for energy reasons, takes about $10^{-11}\,\mathrm{s}$, and during this time up to 35–40 % of the muons in bound states of the $\mu\mathrm{He}^+$ ion after the fusion reaction (14.21), are freed for catalysis. At lower densities, in gaseous mixtures, the fraction of such muons – the so-called *reactivation coefficient* R – is somewhat lower.

The resultant loss of muons for catalysis with account taken of both its sticking to the α particle and its reactivation in the slowing-down process of the $\mu\mathrm{He}^+$ ion is calculated by the formula

$$\omega_{\mathrm{s}} = \omega_{\mathrm{s}}^0 (1 - R)\,, \tag{14.25}$$

where ω_s^0 is the initial, and ω_s the final, sticking coefficient. According to recent calculations the initial sticking coefficient, ω_s^0, in the dtμ fusion reaction amounts to 0.88–0.89 %. With $R = 0.30$–0.35 this gives for the final sticking coefficient $\omega_{s,theor} = 0.57$–0.62 %.

The sticking coefficient is measured by three methods: (a) by the neutron yield; (b) by detection of the charged products of nuclear fusion; (c) by the yield of characteristic X-ray radiation, resulting from population of the $2p$ state and, also, of higher states of the μHe^+ mesoion during its slowing down. Although the ω_s values obtained by different groups and by different methods still disagree, most data for dtμ fusion fall into the interval $\omega_{s,exp} = 0.4$–0.5 %. Thus, from the point of view of the resultant catalysis efficiency the actual situation looks even better than predicted by modern theory.

The search for conditions under which the reactivation coefficient R can be somehow enhanced, and ω_s correspondingly reduced, continues. To this end theoretical studies into the possibilities of μ catalysis in a plasma medium are under way. Versions are being elaborated upon for the enhancement of muon reactivation by applying additional accelerating fields to influence the μHe^+ ion being produced. In connection with the tempting prospects of applying the muon-catalyzed fusion in nuclear engineering, a large number of new and interesting technical issues have been raised. It must be taken into account that an energy release of the order of 2.5 GeV, corresponding to the value of 150 achieved for the number of cycles per muon, is far from providing a solution for the problem by itself, since the production of a single muon at modern accelerators requires not less than 8 GeV. Hopes are due to theoretical developments concerning "hybrid" systems, in which a μ-catalytic reactor is supposed to operate in combination with electro-nuclear breeding. The first predictions reveal that on the whole the efficiency of such a system may be quite high if it is utilized not only for obtaining energy directly, but also for producing fuel for ordinary nuclear reactors.

References to Figures and Tables

Introduction

A.1 Particle Data Group: "Review of Particle Properties"; Phys. Lett. B **239** (1990) 1–516
B.1 Mougey J., et al.: Nucl. Phys. A **262** (1976) 61
B.2 Nakamura K., et al.: Nucl. Phys. A **271** (1976) 221
B.3 Jolos R.B., et al.: Yad. Fiz., **22** (5) (1975) 965
B.4 Speth J., et al.: Phys. Rep. **33** (1977) 127
C.1 Radzig A.A., Smirnov V.M.: *Parameters of Atoms and Atomic Ions* (Energoatomizdat, Moscow, 1986)

Lecture 2

2.1 Gott Yu.B.: *Interaction of Particles with Matter in Plasma Studies* (Atomizdat, Moscow, 1978)

Lecture 3

3.1 Gras-Marti A.: Nucl. Instrum. Methods B **9** (1985) 1

Lecture 4

4.1 Albat R., Bruen N.: J. Phys. B **8** (1975) 959
4.2 Xia Yue-yuan, Tan Chun-yu: Nucl. Instrum. Methods B **13** (1986) 100
4.3 Andersen L.H., et al.: Phys. Rev. Lett. **57** (1986) 214
4.4 Ashlay J.C., et al.: Phys. Rev. **85** (1972) 2393
4.5 Zaikov V.P., et al.: VINITI 600–187 (1987) 75
4.6 Brown M.D., Mool C.D.: Phys. Rev. B **6** (1972) 90

Lecture 5

5.1 Robinson M.T., Oen O.S.: Phys. Rev. **132** (1963) 2385
5.2 Jin H.S., Gibson W.M.: Nucl. Instrum. Methods B **13** (1986) 76
5.3 Miller P.D., et al.: Nucl. Instrum. Methods B **13** (1986) 56
5.4 Andersen S.K., et al.: Nucl. Phys. B **144** (1978) 1
5.5 Tulinov A.F.: Vestnik Mosk. universiteta. Series of physics, astronomy, 1967, No. 5, p. 88
5.6 Domeji B., Bjorkvist K.: Phys. Lett. **14** (1965) 127
5.7 Tulinov A.F.: Vestnik Mosk. universiteta. Series of physics, astronomy, 1967, No. 5, p. 88
5.8 Sigle W., et al.: Nucl. Instrum. Methods B **2** (1984) 1

Lecture 6

6.1 Kolos W., Peak J.M.: Chem. Phys. **12** (1976) 381
6.2 Gemmel D.S.: Chem. Rev. **80** 1980 301
6.3 Parilis E.S., et al.: *Atomic Collisions on Solid Surfaces* (Elsevier, 1993)
6.4 Balashova L.L.: Rad. Effects and Defect in Solids **115** (1991) 289
6.5 Reijnen P.H.F., et al.: Surface Science **221** (1989) 427
6.6 Heiland W.: Vacuum **39** (1989) 367
6.7 Balashova L.L. et al.: J. Phys. Condens. Matter **4** (1992) 4883

Lecture 7

7.1 J. Phys. Chem. Ref. Data. **9** (1980) 1023

Lecture 8

9.1 Komarov F.F., Novikov A.P.: JTF I **48** (1978) 1449
9.2 Andersen J.U.. et al.: Nucl. Instrum. Methods **132** (1976) 507
9.3 Moussa et al.: Physica **40** (1969) 517
9.4 Theodosiou C.: Phys. Rev. A **16** (1977) 2232
9.5 Cairus J.A., et al.: *Atomic Collisions in Solids IV* (Gordon and Breach, New York, 1972) 5
9.6 Agan'yants A.O., et al.: Pis'ma JETP **29** (1979) 554
9.7 Cue N. et al.: Phys. Lett. A **80** (1980) 29
9.8 Swent R.L., et al.: Phys. Rev. Lett. **43** (1979) 1723
9.9 Alguard M.J., et al.: Phys. Rev. Lett. **42** (1979) 1148
9.10 Bazylev V.A., et al.: JETP **80** (1981) 608

Lecture 13

13.1 Hartman F.J., et al.: Phys. Rev. Lett. **37** (1976) 331
13.2 Burbidge G.R., de Borde A.N.: Phys. Rev. **89** (1953) 89
13.3 Yamazaki T., et al.: Phys. Rev. Lett. **63** (1989) 1590

Lecture 14

14.1 Bystritsky V.M., et al.: JETP **76** (1979) 460

Subject Index

Springer
and the
environment

At Springer we firmly believe that an
international science publisher has a
special obligation to the environment,
and our corporate policies consistently
reflect this conviction.
We also expect our business partners –
paper mills, printers, packaging
manufacturers, etc. – to commit
themselves to using materials and
production processes that do not harm
the environment. The paper in this
book is made from low- or no-chlorine
pulp and is acid free, in conformance
with international standards for paper
permanency.

 Springer